Simulation Interoperability

Challenges in Linking Live, Virtual, and Constructive Systems

Dr. Roger Smith

M

Modelbenders Press

PRINTED IN THE UNITED STATES OF AMERICA

Visit our web site at www.modelbenders.com

Designed by Adina Cucicov at Flamingo Designs

The Library of Congress has cataloged the paperback edition as follows:

Smith, Roger
 Simulation Interoperability: Challenges in Linking Live, Virtual, and
 Constructive Systems.
 Roger Smith—3rd ed.
 1. Simulation   2. Computer Science   3. Training and Education
 I. Roger Smith   II. Title

ISBN-13: 978-0-9823040-5-1
ISBN-10: 0-9823040-5-6

Foreword

This book was originally written in 1994. The interoperability examples and state of the art described in the text reflect that period. Its value today is in presenting the evolution of interoperability and providing a historical perspective on progress that has been made in the last 15 years.

The United States Department of Defense has invested heavily in computer simulations since World War II. As a result, simulations and models exist to replicate or evaluate most combat systems and situations. One large class of simulations consists of those used for training personnel to perform their operations more successfully. These are recognized by the general public as "war games" and "flight simulators".

As training simulations evolved, a natural dichotomy of functionality and interface emerged which lead to major differences between: Constructive, Virtual, and Live simulation. Horizontal and vertical interfacing of these classes of simulation is an important step in the future development of training systems. It is also seen as a means of decreasing the cost of future development. Rather than creating multiple models of the same function, as has been done in the past, it will be possible to create one that complies with the interface standards. This will then operate in conjunction with any other model that also meets these interface standards.

In this book, we will explore methods for integrating constructive and virtual level simulations. The virtual community is beginning to develop semi-automated forces (SAFOR) algorithms that will allow them to join with constructive simulations. Enhanced versions of SAFOR will be the bridge that connects the constructive and virtual worlds. These will perform the fidelity enhancement and removal needed for two-way communications across the vertical bridge.

The simple broadcast backbone networks used in both the constructive and virtual communities will not support the types of exercises envisioned for the military. A more complex structure must be imposed on the network, something similar to the Internet. Each node will simulate events in a specified geographic area. Within a node or LAN, these areas may be further subdivided into a military command structure.

Table of Contents

List of Figures 3
Chapter 1. Introduction 5
1.1 Background 5
1.2 Origin of the Models 8
   1.2.1 Constructive Simulations 9
   1.2.2 Virtual Simulators 11
   1.2.3 Live Simulation 13
1.3 Interoperability 13
1.4 Enabling Technologies 15
1.5 Contribution of this Work 17
1.6 Organization 20
Chapter 2. Evolution of Simulation Interoperability 21
2.1 Incorporation of Models within Simulations 23
2.2 Independent Processes on a Single Machine 28
2.3 Simulations with Distributed Components 32
   2.3.1 Local Area Networks 33
   2.3.2 Wide Area Networks 36
2.4 Dedicated Horizontal Interfaces 43
   2.4.1 CBS—AWSIM Interface 43
   2.4.2 CBS—CSSTSS Interface 47
   2.4.3 CBS—TACSIM Interface 48
2.5 Parallel Simulation Techniques 50
2.6 Aggregate Level Simulation Protocol 57
   2.6.1 Origins 57
   2.6.2 Operations 58
2.7 Distributed Interactive Simulation 66
2.8 Computer Generated Forces Models 77
   2.8.1 ModSAF 79
   2.8.2 IST SAFOR 84
   2.8.3 IBM Blackboard 84
   2.8.4 Rasputin 84
2.9 Vertical Integration Prototypes 86
   2.9.1 AWSIM to ModSAF 86
   2.9.2 JPL Alpha Project 88
   2.9.3 Eagle to BDS-D/SIMNET 89
   2.9.4 BBS to SIMNET 93
Chapter 3. Translational Functions 99
3.1 Relationships 100
   3.1.1 Templates 100
   3.1.2 Dynamics 102
   3.1.3 Consistent History 104

3.1.4 Control 106
3.2 Actions 107
3.2.1 Movement 107
3.2.2 Detection 110
3.2.3 Communications 113
3.2.4 Posture 114
3.2.5 Engagement—Direct 115
3.2.6 Engagement—Indirect 117
3.3 Timing 119
3.4 Disaggregation 121
3.4.1 Units 121
3.4.2 Ordnance 124
3.4.3 Consumption 127
3.4.4 Orders 129
3.4.5 Firewalls 130
3.5 Aggregation 133
3.5.1 Units 133
3.5.2 Ordinance 134
3.5.3 Consumption 135
3.5.4 Orders 135
Chapter 4. Organizational Architectures 137
4.1 Network Architecture 138
4.1.1 Backbone 138
4.1.2 Internet 138
4.2 Organizational Alternatives 139
4.2.1 Model Level 140
4.2.2 Military Unit Level 141
4.2.3 Military Command Level 142
4.2.4 Functional Level 143
4.2.5 Geographic Areas 144
4.3 Computer Generated Forces 145
4.3.1 Virtual 145
4.3.2 Constructive 145
4.3.3 Future 146
4.4 Prototype Organizations 147
4.5 Preferred Organization 148
4.5.1 Pure Geographic 148
4.5.2 Geographic/Command Hybrid 148
Chapter 5. Recommendation for Future Work 149
Bibliography 151
Appendix A—Acronyms 169
Appendix B—Glossary 173

List of Figures

Figure 1-1. [Singley, 1993]..5
Figure 1-2. Conceptual Architecture for a Vertically
    Integrated Training Simulation................................6
Figure 1-3. One View of Relationships Between the
    Different Simulation Levels......................................14
Figure 1-4. ALSP/DIS Conceptual Interface.................17
Figure 1-5. Vertical Integration Tasks............................18
Figure 1-6. Interface Organizational Architecture........18
Figure 2-1. Interoperability Evolution Tree...................22
Figure 2-2. Integration of High Fidelity Model.............27
Figure 2-3. Database Oriented Simulation Structure....30
Figure 2-4. Training Simulation Operational Environment...33
Figure 2-5. Sample Simulation Exercise WAN...............36
Figure 2-6. Simulation Data Packet................................37
Figure 2-7. CBS Workstation Connections.....................38
Figure 2-8. RPC Structure...42
Figure 2-9. Logical Processes in Parallel Simulation.....51
Figure 2-10. Aggregate Level Simulation Protocol Design...59
Figure 2-11. DIS Architecture..69
Figure 2-12. DIS Functional Requirements....................70
Figure 2-13. DIS PDU Requirements..............................71
Figure 2-14. Physical DIS Network Interface.................72
Figure 2-15. DIS Network Performance Analysis...........74
Figure 2-16. Virtual Body Coordinate System...............75
Figure 2-17. ModSAF Development Timeline.................79
Figure 2-18. ModSAF Components.................................81
Figure 2-19. ModSAF Model Libraries...........................82
Figure 2-20. AWSIM to ModSAF Functionality.............87
Figure 2-21. JPL Project Alpha Links.............................88
Figure 2-22. Eagle to BDS-D Functionality....................90
Figure 2-23. Eagle to IST SAFOR Gateways..................91
Figure 3-1. Constructive-Virtual Unit Relationships....100
Figure 3-2. Simple Unit Movement................................107
Figure 3-3. Complex Unit Movement.............................110
Figure 3-4. Simple Unit Detection..................................111
Figure 3-5. Layers of Disaggregation.............................121
Figure 3-6. CBS-TACSIM Disaggregation Template.....122
Figure 3-7. Interaction of Objective and Environment...130
Figure 4-1. Model Level...140
Figure 4-2. Military Unit Level.......................................141
Figure 4-3. Military Command Level..............................142
Figure 4-4. Operational Level...143
Figure 4-5. Geographic Areas...144

Introduction

1 1 0 0 1 0 0 0 1 1 0 0 1 0 1 0 0 0 1 1 0 0 1 0 1 0 1 0 1 0 0 0 1 1 0 0 1 0 1 1 1 0 0 1 1 1 0 0 1 0 1 0

1.1 Background

The United States Department of Defense has invested heavily in computer simulations since World War II. As a result, simulations and models exist to replicate or evaluate most combat systems and situations. One large class of simulations consists of those used for training personnel to perform their operations more successfully. These are recognized by the general public as "war games" and "flight simulators".

As training simulations evolved, a natural dichotomy of functionality and interface emerged which lead to major differences between:

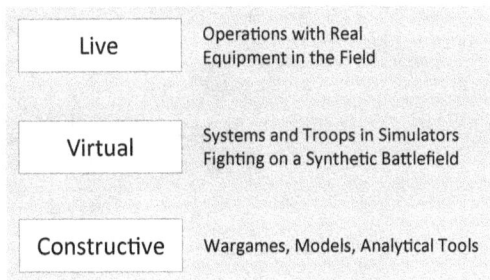

Live	Operations with Real Equipment in the Field
Virtual	Systems and Troops in Simulators Fighting on a Synthetic Battlefield
Constructive	Wargames, Models, Analytical Tools

Figure 1-1. [Singley, 1993]

- Constructive Level (war games),
- Virtual Level (flight simulators), and
- Live (target and flight ranges) simulation (Figure 1-1) [Singley, 1993]. After nearly 40 years of progress, these three sub-communities have grown so far apart that they do not and can not interoperate. As a result, each group can not take advantage of the progress made by the other two, often resulting in duplicated work.

Horizontal and vertical interfacing is seen as an important step in the future development of military simulations. Many benefits ensue. It will be possible to create a more complete and accurate environment in which to train our defense forces by joining these together. It is also seen as a means of decreasing the cost of future development. Rather than creating multiple models of the same function, as has been done in the past, it will be possible to create one which complies with the interface standards.This will then operate in conjunction with any other model which also meets these interface standards (Figure 1-2).

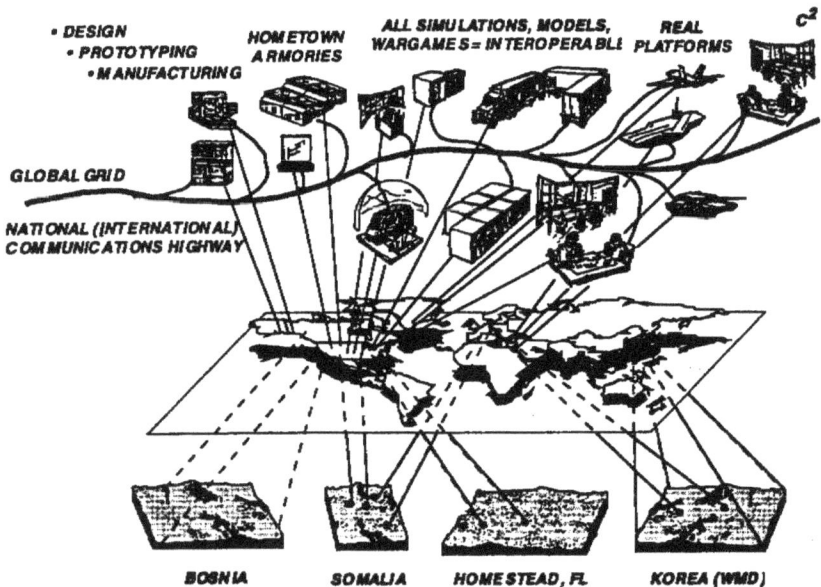

Figure 1-2. Conceptual Architecture
for a Vertically Integrated Training Simulation

The 1991 Defense Science Board studied the problem of the lack of interoperability between these communities and between different simulators within each community

[DMSO, 1993]. Their recommendation was that projects be initiated to link all simulations and allow them to participate together in a single simulation environment. In response, the Virtual Level community created the Distributed Interactive Simulation project (DIS). The Constructive Level community created the Aggregate Level Simulation Protocol project (ALSP).

The Live simulation community faces a much more difficult task. Their simulators are actually the aircraft and vehicles of war. Interfacing these requires a built-in simulation mode which can be switched <u>on</u> when engaged in an exercise, and <u>off</u> when engaged in actual operations. These ideas are being included in new procurements, but some equipment can currently be interfaced with simulations due to the design of the simulation. The intent in the Live world is to use the DIS standards. Hence, solving the ALSP to DIS mergence problem will provide a large measure of the connection needed to include the Live community.

Both the ALSP and DIS projects have made serious progress within their own environs, but have made little effort to join with one another. The number of problems to be solved by each project is so vast, the impact so far reaching, and man-power so limited, that it is difficult to interest either community in this vertical integration task. However, the Department of Defense has continued to emphasize its interest in integrating all simulations, regardless of their level of fidelity. As a result, a few projects are underway to bridge this gap. These researchers have begun by addressing minor problems and attempting to demonstrate special case integration, rather than completing the entire vertical integration project at once.

1.2 Origin of the Models

Now that the government is so focused on connectivity, it is easy to criticize earlier models for not catering to this need. But, an understanding of the origin and mission of existing models will be helpful in understanding their capabilities and future needs.

Wargaming is an ancient art. As long as there have been organized wars there has been some form of wargaming. Drawing in the sand and moving stones as forces against the enemy may have been the beginning. Originally used to plan an attack, the techniques became a means of re-telling a battle and then a form of training. Playing out a war in the sand with markers representing friendly and enemy forces was used to teach lessons learned in past battles. It also became a means of exploring "what if" scenarios; enabling visualization of tactics and inviting reactions from others.

During the Roman era, wargaming was formalized into "sand tables" and unit icons. A large sand box was land-scaped into a representation of the field of battle. Trees, hills, boulders, cliffs, rivers, and swamps were all repre-sented. Military forces were simulated with wooden or stone carvings of men, horses, and tents. Two command-ers could then face each other across the sand and explore the possible outcomes of planned tactics. As a training tool this was, and still is, invaluable. In fact, modern sand tables that fill entire gymnasiums can be found in use by the Army in places like Fort Leavenworth, Kansas and Fort Hood, Texas. While soldiers may train with their weapons in drills, commanders use wargames to exercise their pri-mary weapon—the mind.

1.2.1 Constructive Simulations

There are hundreds of models within the constructive community. These are usually referred to as constructive models, since individual objects (tanks, helicopters, etc.) are grouped into operational constructs such as armor companies and air defense batteries. In this paper we limit ourselves to the most widely used and accepted of these. Hundreds of models exist because analysts with a problem to solve often construct a unique model of it, complete their study, then put it aside until it is needed again. There has also been a tendency for projects to build their own new models, rather than adapting one built by others. The largest and most widely accepted of these are:

Corps Battle Simulation (CBS)—This model is used to train the decision making abilities of corps, division, and battalion commanders. It focuses on troop movement, attrition, artillery fire, and follow-on-force employment.

Brigade/Battalion Simulation (BBS)—This model is a smaller scale model following in the footsteps of CBS, but meant specifically for battalion commanders.

Air Warfare Simulation (AWSIM)—This model is used by the US and NATO Air Forces to train wing and squadron commanders in the employment of their air assets. It focuses on air-to-air, air-to-surface, and surface-to-air combat. It models air movement, attrition, electronic warfare, air refueling, ground strikes, and the response against these aircraft by surface-to-air missiles.

Tactical Simulation System (TACSIM)—This model simulates the performance of division and theater intelligence collection assets. It is meant to train the intelligence analysts that provide enemy situation maps to the combat commanders.

Joint Electronic Combat / Electronic Warfare Simulation (JECEWSI)—This model focuses on the electronic warfare environment. It includes the use of air- and ground-based radio frequency (RF) jammers, and radar collection and detection.

Combat Service Support Training Simulation System (CSSTSS)—This models the use of the logistic capabilities to provide supplies to forces in combat. Supplies modeled include ammunition, missiles, food, fuel—almost everything necessary to equip and maintain an army in combat.

Battle Intelligence Collection Model (BICM)—This is an intelligence collection model meant to bypass the intelligence analysts and provide enemy situations in an analyzed form directly to the combat commanders.

Aggregate Level Simulation Protocol (ALSP)—This is a family of software and communication protocols which join all of the above models into a loose confederation, operating together.

Before the advent of ALSP, models were connected on a one-to-one basis. When it became necessary to connect the air and ground war, an interface was built to transfer data between CBS and AWSIM. This interface was designed with one mission in mind and no thought given to generality or reuse. Therefore, the software could only be used for this one connection. The trend having been set, other interfaces were soon needed to bring the strengths of each model into a single exercise environment. Interfaces were soon built between CBS-TACSIM, CBS-JECEWSI, CBS-BICM, etc.

Once the benefits of this type of distributed modeling were experienced, it became desirable to construct a generic interface standard which could be used for all current and future models. This emerged as ALSP.

1.2.2 Virtual Simulators

Interfacing at the Virtual level emerged a bit differently. Simulators of fighter aircraft and armored vehicles were developed to train the equipment operators without actually using the equipment. This soon spread to commercial aircraft and industrial vehicles. The first large scale integration project was sponsored by the Defense Advanced Research Projects Agency (DARPA). A group of tank simulators were joined to form a family known as the Simulator Networking (SIMNET) project. This soon became renowned for its ability to immerse multiple people and teams in a single virtual combat environment. Tank crews were able to fight against each other, even to the point of firing their weapons, without risking each other's lives. Following this success, DARPA (now ARPA) undertook to include flight simulators, helicopters, armored vehicles, and even infantry into this virtual environment. This project is the Distributed Interactive Simulation (DIS). The focus is much the same as ALSP, but at the virtual level.

On the DIS side, SIMNET is being replaced with ever more complicated simulators under the Army's Combined Arms Tactical Trainer (CATT) program. These along with a host of existing Virtual level simulations, will make up the DIS network. The CATT program includes:

Close Combat Tactical Trainer (CCTT)—This is the direct follow-on to SIMNET. It replicates the functions of armored and mechanized vehicles such as the M1A1 tank, M1A2 tank, and the Bradley Fighting Vehicle.

Aviation CATT (AVCATT)—This will replicate Army air assets, primarily Apache and Cobra Attack helicopters.

Air Defense CATT (ADCATT)—Simulates the performance of air defense batteries such as the Patriot.

Field Artillery CATT (FACATT)—Simulates operations of 155mm artillery and Multiple Launch Rocket Systems (MLRS).

Engineering CATT (ENCATT) - Simulates the operations of engineering vehicles such as mine layers/removers, bulldozers, and bridging equipment.

These simulators will be tested under a program known as Battlefield Distributed Simulation-Developmental (BDS-D). The experiments will focus on operations: Line-of-sight Anti-tank (LOSAT), Command Ground Station (CGS), Combined Arms Command and Control (CAC2), Combat Identification (CID), and Rotorcraft Pilot's Associate (RPA). All of these focus on testing the capabilities of the simulators to provide realistic, integrated training in the virtual environment [Singley, 1993].

1.2.3 Live Simulation

The third form of simulation identified by the 1991 Defense Science Board is Live Simulation. This is the operation of fighters in training exercises as popularized by movies such as "Top Gun". Although operating the actual equipment the activity is a simulation in that it involves staged combat between the aircraft. Exercises are carried out by the Air Force in the "Red Flag" program, the Navy "Top Gun" program, the Army "Green Flag", and a host of others.

1.3 Interoperability

As computer and networking technology has matured the ability to replicate reality within them has increased. We are now at a point where we can join thousands of these simulators into a single environment and create events which are useful in a training sense, and valuable from a research point of view.

One major problem that has arisen is the fact that ALSP models update unit activities and status every 1 to 6 minutes, while DIS models do so every 100 to 300 milliseconds, a 1:600 difference [Smith, 1992 and The DIS Vision, 1993]. The difference is obviously due to the level of fidelity of object performance. A flight simulator pilot expects to see the terrain and other objects move past him in real time. But the division commander is used to watching movement in terms of hours. Obviously some bridge or translation mechanism is needed to remove details from the data flowing up to ALSP and add it to data flowing down to DIS.

One bridge has evolved from a project designed to provide automated but reactive opponents in the DIS world. In a SIMNET exercise, there is a definite need for more friendly and enemy objects than there are simulators and crews to man them. Without this, the virtual world consists of a handful of tanks hunting each other. But, realism may require hundreds or thousands of vehicles all engaged in independent war fighting activities. To provide this level of interaction, the SIMNET project developed the Semi-Automated Forces model (SAFOR). This is responsible for creating intelligent, reactive, and realistic fighting forces to join with or fight against the tanks actually manned by human crews. Thus far, it has succeeded at only a very crude level of realism, but it is the beginning of more advanced techniques. Generally, the concept has been dubbed Computer Generated Forces (CGF) and is being pursued by several different offices (Figure 1-3).

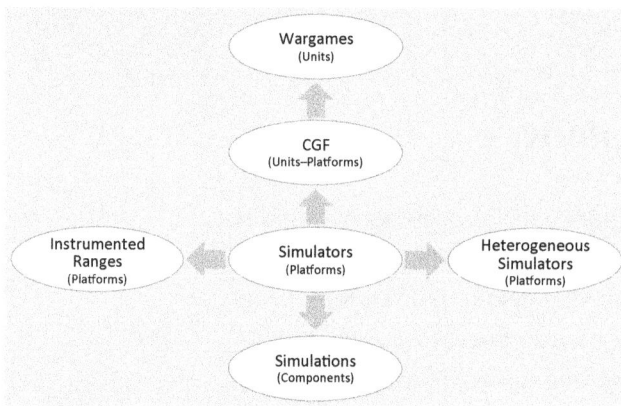

Figure 1-3. One View of Relationships Between the Different Simulation Levels

Since Computer Generated Forces use state and event information to create a realistic force, they are performing some of the functionality needed to add fidelity to information flowing down from ALSP to DIS. This is currently being explored as a bridging technique. By reversing the process, CGF may also operate as a filter, removing details from data flowing from DIS to ALSP.

Given that such bridges are necessary, we will explore the types of functions that must be incorporated into them. We plan to show which details in the DIS-world must cross into ALSP-space, and which do not. We will also show the effects of playing aggregated units in ALSP, while smaller parts of these are played as manned objects in the DIS-world.

1.4 Enabling Technologies

Earlier we referred to the recent advancements in technology which have made simulation possible and desirable. Here we will briefly outline these technologies and their contributions.

Power. The first is the extreme improvement in computer processing power. The power of early computers had a definite sizing effect on the simulations developed on them. Each model was seen as being the size of the host computer, usually a DEC VAX. As distributed computing has come to the forefront this limitation is being shattered.

Networking. Distribution via networking has seriously altered the mission of combat simulations. Where CBS, TACSIM, etc., were originally used to train a local audience, they are now networked together and operate across the globe. In fact, it is not unusual for the computer running a simulation in Germany to be physically located in Kansas. The Blue forces may be operating in tents in the fields outside small German villages, but their opponents are at the National Simulation Center, Fort Leavenworth, Kansas. Time zones and geography are irrelevant, as both groups are immersed in the simulated events inside the computer. At first, computers were joined in a Local Area Network in which the communication medium was a dedicated line wired between the computers. Later, Wide Area Networks where assembled, and then enhanced by the advent of fiber optics and communications satellites.

Graphics. Graphics have had a profound impact on the utility of computer simulation. Early models operated with input and output in textual format. It was then up to the humans to translate this into graphic form, usually as tactical maps of unit locations and terrain. Today, the computer also controls these displays and automatically updates them as the simulation progresses. Unfortunately, graphics have not been implemented as an input form, as much as for output. This is primarily due to the difficulty in creating a user input interface which is quicker, more flexible, and more intuitive to use than a keyboard.

Modeling. The last advancement is the development of military combat simulation and modeling techniques. These are noticed only by those in the industry, but they have been significant. Unit representation in TACSIM has enabled it to model the operations of thousands of units, containing tens of thousands of objects on a very limited computer budget. Monte Carlo modeling is a relatively new field. As with computers one of its first proponents was John Von Neumann in the 1950's. The field developed quickly into the 1980's, but is now experiencing a plateau, primarily due to the attention being focused on new computing technologies.

1.5 Contribution of this Work

We will attempt to explore and suggest some solutions to
the vertical integration problem in this paper. Specifically,
we will be looking at methods for managing the changes
in fidelity required by units at the constructive level and
those at the virtual level. This must be done in an
environment in which several models are interoperating
at both levels. Therefore, the task goes beyond a simple
generic connection between two models. Interactions and
information cross talk must be considered (Figure 1-4).
When one tank transfers its position up to the constructive world, it becomes more than a single message. The

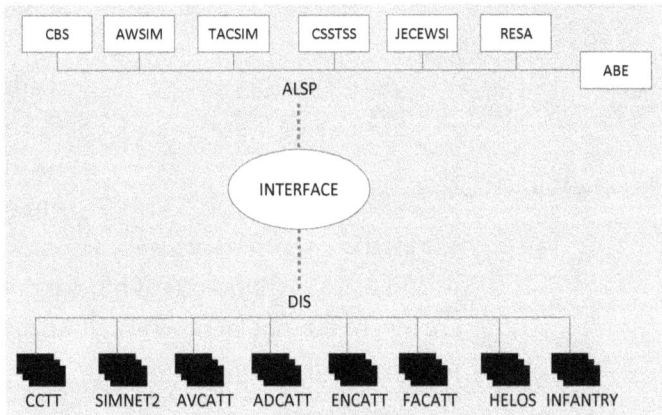

Figure 1-4. ALSP/DIS Conceptual Interface

information must go to all constructive models concerned
with the operation of the organizational unit owning that
tank, which may number from one to one hundred.
Therefore, all models of interaction must scale well,
preferably $O(\log n)$.

On the other hand, a method is needed for adding detail
to unit commands flowing down into the virtual level.
Although an aggregated unit may be told to move from
point A to point B in a generally straight line, its virtual
counterpart must translate this into a specific path
through detailed terrain representations. It must also

account for and react to any intervention by obstacles such as missing bridges and encounters with enemy units. The effects of these details must then be reflected in the ope-rations of the aggre-gated unit in the upper levels (Figure 1-5).

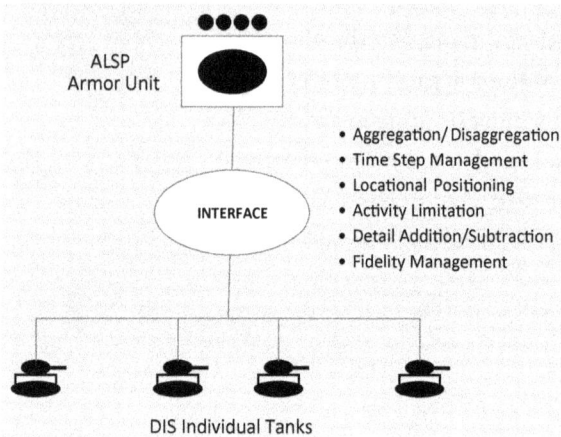

Figure 1-5. Vertical Integration Tasks

- Aggregation/Disaggregation
- Time Step Management
- Locational Positioning
- Activity Limitation
- Detail Addition/Subtraction
- Fidelity Management

These problems are not trivial, which is one reason for the slow progress being made in connecting the two worlds. But, neither is the task impossible. In the real world, these translations are made constantly during wartime, where hundreds of human minds and bodies are involved in the process which we intend to accom-plish with a few computers.

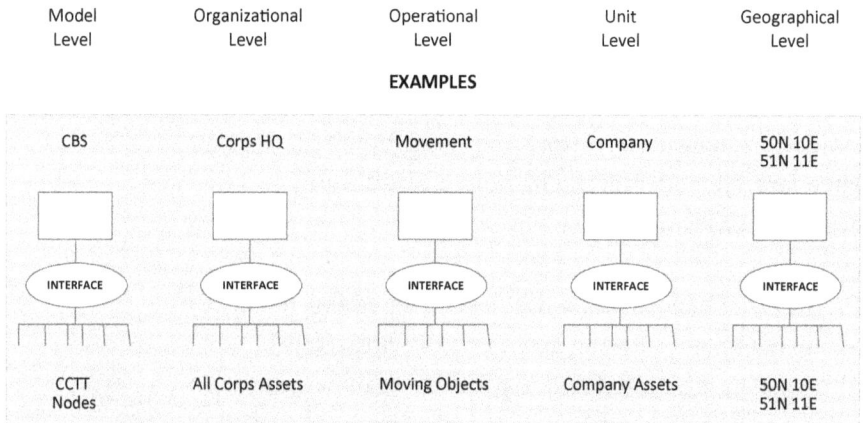

Figure 1-6. Interface Organizational Architecture

If we are to separate the interface into functional parts, rather than use one large conduit, we must select a method of organization. Several options are workable (Figure 1-6). All interactions with a given constructive level simulation may be assigned a communication channel and interface device. Grouping may be done at the organizational command level. Here, all information concerning units or objects belonging to an operational command headquarters use the same interface. Different types of unit operations may share an interface unit. Here, all movement messages use one conduit, resupply messages another, and engagement messages still another. Finally, information may be grouped geographically. All messages originating or destined for a certain area may be assigned the same interface channel.

All of these methods will be examined in this paper, along with the types of information exchanged in the messages. Additionally, we will explore methods for translating the information so that it is appropriate to the level of the intended receiver.

1.6 Organization

This book is organized into five chapters. Chapter 1 is an introduction to the problem and the current simulation community. Chapter 2 is a review of the simulation literature and earlier work that has been done on the vertical integration problem. Chapter 3 describes the types of translations that need to be performed on data traveling across the vertical bridge. These translations will be performed by interfacing models which are inserted between the ALSP and DIS networks. Chapter 4 describes possible organizational architectures for transmitting data between constructive and virtual simulations. This describes the functional architecture for organizing data transfer. It does not discuss the physical connectivity needed to carry the data. Chapter 5 provides a summary of the findings of the paper and recommends avenues of future research on the vertical integration problem.

Evolution of Simulation Interoperability

110010001100101000110010101010001100101110011100101 0

Since vertical integration is in such an infant state, there is little literature available in the form of textbooks. Most information is in the form of conference proceedings and meetings between simulation proponents. These include semi-annual workshops on designing the DIS interface; quarterly meetings to design the ALSP interface; conversations between members of the community; existing interfaces between specific simulations; annual conferences sponsored by the Society for Computer Simulation, IEEE, and the Association for Computing Machinery; and, finally, electronic conferences on simulation.

This chapter will examine several of the existing and experimental methods for interfacing simulations. These include:

- Incorporation of Models within Simulations,
- Independent Processes on a Single Machine,
- Simulations with Distributed Components,
- Dedicated Horizontal Interfaces,
- Parallel Simulation Coordination,
- Aggregate Level Simulation Protocol,
- Distributed Interactive Simulation,
- Computer Generated Forces Models, and
- Vertical Integration Prototypes.

The rest of this chapter will be devoted to exploring these nine interface types. Special attention will be given to the aspects of these which are pertinent to the topic of this book. For this reason, the sections are not exhaustive, more information on each can be found in the sources listed in the bibliography. The dynamic nature of the vertical prototypes insures that any attempt to catalog their operations will necessarily be limited to the capability available at the time of writing. All of these will have progressed by the time this book is published. This book attempts to make a contribution to the evolution of interoperability by exploring ideas for vertical integration standards which will hopefully lead to full, seamless integration of all simulations (Figure 2-1).

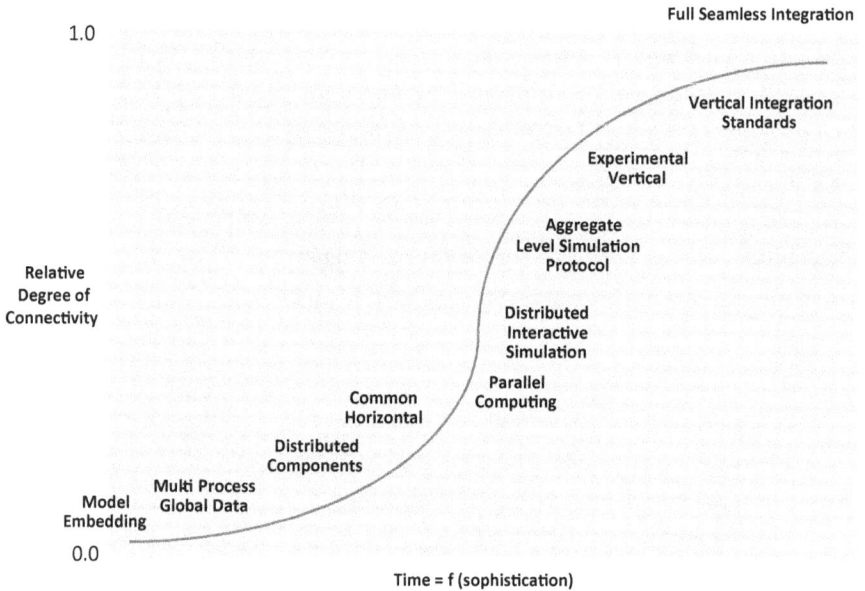

Figure 2-1. Interoperability Evolution Tree

2.1 Incorporation of Models within Simulations

A basic method of simulation integration is accomplished by embedding one model inside another. This is usually done by allowing a specialized, high-fidelity model of some sub-system of the larger system to handle all events pertaining to its area of expertise.

We shall explain this through the use of examples taken from the ALBAM simulation. ALBAM was built to simulate the entire theater of war. This includes events in direct-fire ground combat, indirect-fire ground combat, logistics operations, combat engineering, ground and airborne electronic warfare, intelligence collection and dissemination, air-to-air combat, surface-to-air engagements, air-to-surface strikes, air-to-air refueling, airbase sortie generation, ground movement across terrain and via road networks, and all of the simulation management, input and output associated with these operations. Excluding the graphics displays and after action data collection, all of these functions are accomplished on a single computer using a single global database [Smith, 1992].

Although the entire theater was replicated, certain areas were identified as highly significant operations. In these cases, a finer level of detail was necessary than could be accomplished for the entire model within the hardware, data, man-power, and schedule constraints. Enhancements were requested for the following areas:

- aircraft fuel consumption,
- aircraft maintenance cycles following missions,
- air-to-air engagements,

- air-to-surface engagements, and
- surface-to-air engagements.

To accommodate the first two on the list, detailed algorithms were incorporated into the ALBAM model. But for the engagements, it was decided that existing specialized models would be integrated into the larger ALBAM model. The specialized models chosen for these were:

- TAC Thunder,
- Terminal Surface-to-Air Missile Model (TSAM), and
- Joint Munitions Effectiveness Models (JMEM).

With an increase in fidelity comes an increase is CPU cost. Therefore, these improvements had to be managed such that the entire ALBAM model was not adversely affected in its ability to process events in real-time. To accomplish this, two methods of integration were selected. TAC Thunder and TSAM would be used to generate multi-dimensional tables of the probability-of-kill (Pk) for a given set of situations. This would be done for a wide range of conditions, encompassing all events that would actually occur during an ALBAM simulation. Generating the tables was a massive undertaking involving hundreds of hours of planning and execution time.

Under this method, when ALBAM determined that an air-to-air engagement was occurring, the variables that govern the engagement were used as indices into the stored TAC Thunder tables to acquire an appropriate Pk. The variables are not exactly those used to generate the tables. In these cases interpolation is done between the surrounding table values.

As an example, the variables provided by ALBAM for each side of the engagement may be:

Cell Size = CS
Speed = S
Altitude = A
Radar = R
Counter Measures = CM
Missile Armament = Mi, Mj, ...

Notice than a group of aircraft may carry more than one type of air-to-air missile. TAC Thunder is built to handle complex engagements between differing weapon types, simultaneously. This is a better method than generating multiple independent engagements for each type of missile, and then trying to introduce dependence in these events by adjusting the resulting probabilities.

The model would then operate on these variables as:

$$f\,[\,R(CS, S, A, R, CM, Mi, Mj, ...)\,,\,\,B(CS, S, A, R, CM, Mi, Mj, ...)\,]$$
$$=\,\,Rk \text{ and } Bk, Rmix, Rmjx, ..., Bmix, Bmjx, ...$$

where,

R(. . .) = Characteristics of Red/Enemy Aircraft Cell,
B(. . .) = Characteristics of Blue/Friendly Aircraft Cell,

Rk = Aircraft Killed in Red Cell,
Bk = Aircraft Killed in Blue Cell,
Rmix = Red Missiles of Type i Expended,
Rmjx = Red Missiles of Type j Expended,

.

.

.

Bmix = Blue Missiles of Type i Expended,
Bmjx = Blue Missiles of Type j Expended,

.

.

.

In the case of the TSAM model, the same method was used but the answer returned was a probability of kill for aircraft in the cell. Monte Carlo methods are then used to determine the actual attrition experienced.

One fact should be obvious as we described the independent variables in the function above. A high-fidelity model requires more detailed inputs in order to produce more accurate outputs. When models like TAC Thunder are run independently, the number of inputs is in the hundreds. These include variables such as cloud cover, wind speeds and directions, aircraft weight—both fuel empty and fuel full, radar parameters, missile seeker-head characteristics, opening aspect angle, sun angle, etc. To join with a model such as ALBAM, these must be set to reasonable values and held constant through all replications used to generate the data tables. In doing this, certain assumptions must be made and recorded. The outcome of the air-to-air engagements are then known to be accurate, subject to these assumptions.

In one sense, these assumptions weaken the fidelity of the TAC Thunder results. It is similar to fighting a battle by using historical texts to present examples similar to the combat situation currently experienced. The selected tactics may not be totally applicable, but the results are more accurate than what could have been replicated without the text.

For the surface-to-air engagements, a different method was selected for integrating the Joint Munitions Effectiveness Models (JMEM). In this case, translation software was built to prepare the variables (similar to those shown in the above example) for input directly into the models. The JMEM software had been integrated into the ALBAM executable binary. As with TAC Thunder,

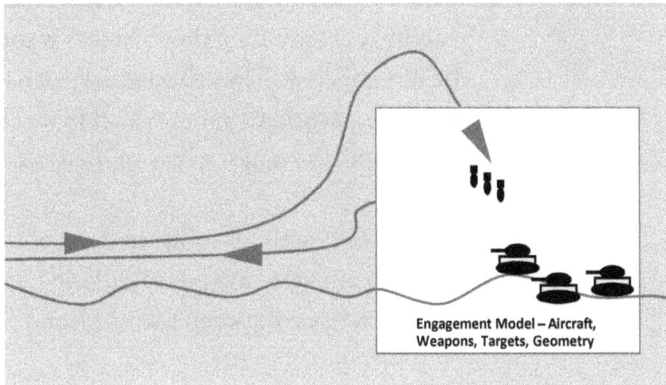

Engagement Model – Aircraft, Weapons, Targets, Geometry

Figure 2-2. Integration of High Fidelity Model

those variables not available had to be set to reasonable values. The computer then executed the JMEM software and returned the results for application by ALBAM (Figure 2-2) [General Dynamics, 1989].

This method does not require the storage space for the calculated tables, but it does increase the size of the simulation software and executable binaries. It is slower than the table method, but provides more accurate results since it does not have to resort to interpolation between pre-calculated values.

It should be pointed out that combat outcomes returned by the high-fidelity models to ALBAM need not be applied directly to the target objects. These are still subject to interpretation or modification within the larger model. In the case of ALBAM, a reliability factor was added to the air-to-surface Pk to produce a reduced kill-level on ALBAM targets. Similar factors are often added for weapon/sensor counter measures and the existing electronic environment. These values represent characteristics not modeled in TAC Thunder or JMEM, and are usually derived from other models or actual combat data. The validity of combining these factors is something that must be investigated. These bridge several data sources and produce a synthetic event which more closely resembles a real-world encounter when applied correctly.

All of these methods indicate some of the steps that are taken to integrate models of differing levels of fidelity, as will be the case between the ALSP and DIS communities.

2.2 Independent Processes on a Single Machine

Although early models were usually limited to a single process operating on a single machine, this paradigm soon gave way to more complex structures. The first step was to expand the operations of the simulation to include multiple models operating on the same machine. The advantage of this on a single processor machine was not in achieving parallelism, but in allowing users' processes to appear to operate in parallel. The advantages of this in a multi-user system are to free the user from the task of waiting for one process to finish in order to execute the next, and to allocate more of the computer's total CPU cycles to the simulation, which now appears as two independent customers on the machine. If these processes were to share data, significant work had to be done to account for the lack of synchronicity between the processes. Once this was accomplished, the most common method for passing this data was to write to and read from files on disk.

The difficulties of operating simulations in this manner are very evident. The modeler must spend much of his/her time overcoming the limitations of the computer's operating system rather than designing and improving the models themselves. Also, simulations of this type do not allow for much human interaction. It was very difficult to input data at appropriate times in a dynamic manner. Since data sharing was done via disk files, human interaction with these can seriously delay or even corrupt the simulation runs. Therefore, this achievement of splitting the model into several pieces was applicable where multiple runs of the simulation could operate independently. But, the next step was to allow separate processes to access the same data pool simultaneously. The primary method of accomplishing this was to create a "global memory" section accessible to any process which requested a mapping to it.

The creation of global or shared memory was very helpful in allowing the user to synchronize multiple processes and observe the operations of the simulation. One process could now be developed to read the shared memory and display its contents. Since files were not being accessed, such processes interfered little with the operation of the models. An I/O process can then allow the operator to play an interactive part in directing the progress of the simulation.

Many models, both simple and complex, were, and are, developed under this structure. Simulations, such as the Air Land Battle Assessment Model, the Tactical Simulation, and the Air Warfare Simulation, operate in this mode today. All three of these are very complex, capable of supporting dozens of users simultaneously. But, excepting the graphic display devices, all of the processes of the simulation are designed to operate on a single computer.

Communications among the different processes that make up the simulation is usually done by passing messages into the global memory area. Each process is assigned locations to look for input and to place output. These data exchange points are in many ways similar to the disk file exchanges described earlier, though they do not suffer from many of the limitations.

In the case of the Air Land Battle Assessment Model (ALBAM), there is a single simulation process which focuses on a shared database (Figure 2-3). Events are scheduled by the simulation itself or by any one of the many command input processes attached to user terminals. The users also have data output processes/ter-

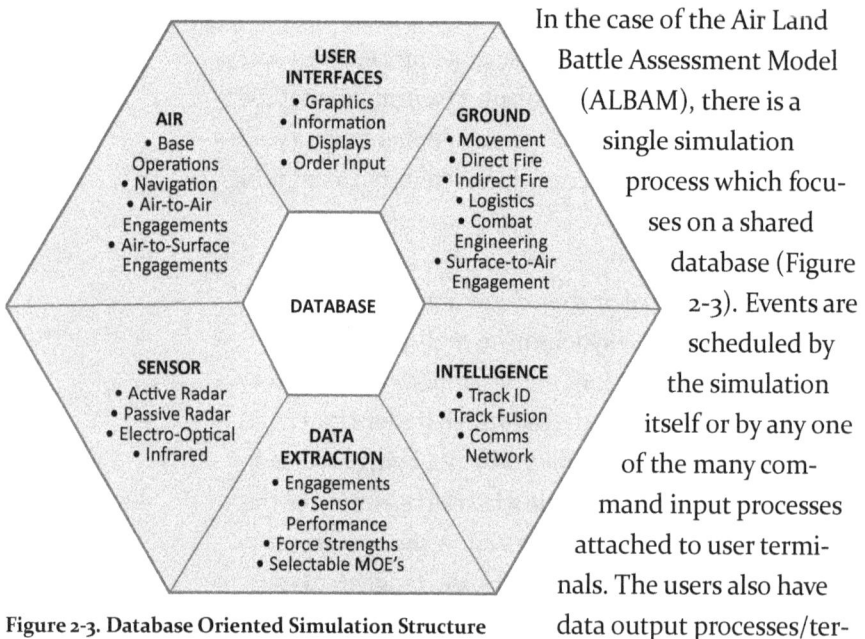

Figure 2-3. Database Oriented Simulation Structure

minals which display tables of information concerning
the progress of the combat engagements being generated.
Time synchronization is not a problem since there is only
one process determining time, and all the others simply
accept it and use it in their operations. The humans operating the simulation are synchronized with this time by
displays on their terminals, which they must accept as the
current time. If this does not flow at the rate expected by
the operators, it may cause disapproval, but they can not
affect any change in it, nor can they cause the execution of
events contradictory to the simulation time.

Originally, tabular combat data was plotted on maps by
a host of clerks charged with the task. When a dynamic
war was unfolding this method became very man-power
intensive and inefficient. A natural alternative was to use
graphic computers to automatically display this data as it
is known. These types of connections are discussed in one
of the following sections.

Since the database is shared there is no need to pass messages between processes. Each process may be responsible
for extracting the information it needs and formatting it
appropriately. Conversely, input processes are free to write
information into the database, as specified in the design.

Models of this type are limited to the power of the
CPU they are tied to. In turn, this is limited by the
state-of-the-art prevailing at the time of the computer's
purchase. Although a simulation may start small, it is
common for additional features to be added until it taxes
the capabilities of the machine. Therefore, all simulations
are limited by the computer, rather than by the scope of
the problem being solved. The model is not scalable.

Interfaces in this structure are purely via the data blackboard constructed in global memory. All exchanges occur according to formats and rules laid out with an in-depth understanding of the operations being performed and the order in which they are performed. Aside from the lack of scalability, the system is limited in that modifications in the method or the order in which data are stored must be done with a complete understanding of the system. Without this it is possible, and probable, that the change will disrupt the operations of another part of the model, either directly or via a trickle-down effect.

2.3 Simulations with Distributed Components

Simulations with multiple processes appear to be natural candidates for distribution among several computers. To accomplish this, hardware, software, and model connections need to be established to maintain the integrity of the simulated world. The connections of interest in this work are the modeling connections. In the examples above, ALBAM was designed to operate with graphics computers producing maps of the locations and types of military units on the battlefield (Figure 2-4). It also had a connection to another computer designed to extract data for combat analysis. These systems presented a simple modeling interface. Both the graphics computers and the after action analysis computer were only reading the data being generated by the simulation, input, and output processes. In this scenario there are no time or event synchronization problems.

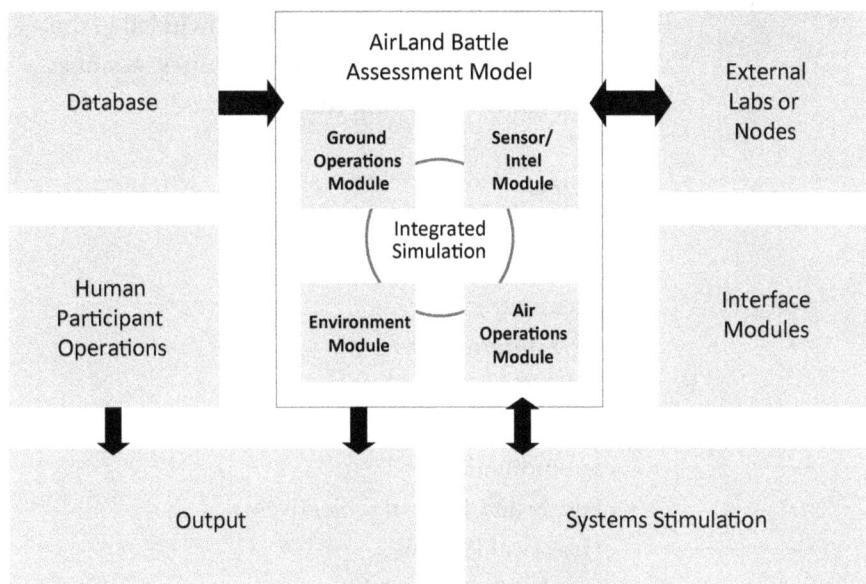

Figure 2-4. Training Simulation Operational Environment

2.3.1 Local Area Networks

Graphics terminals have usually been output devices. As sophistication has grown, these are now able to manipulate the graphics scene and filter the data being presented. But they still enjoyed a one-way interface with the simulation process and its database. In this case, a few simple algorithms added to the simulation process can identify, extract, package, and transfer the data elements needed for the graphics display. Here, the coordination needed to capture a consistent picture of the simulation data is a natural characteristic of the fact that the simulation models stop, while the graphics service routines capture and transfer their data packets [Smith, 1992].

Given that a simulated entity, like a military unit, is assigned hundreds of characteristics, only those effecting the graphics images to be displayed are included in the data packets captured by the service routines. Assume that a unit has the following characteristics:

- Name,
- Size,
- Type,
- Location,
- Orientation,
- Velocity,
- Activity,
- Subordination,
- Future and Current Objectives,
- Historical Records,
- Combat Equipment Lists,
- Sensor Equipment Lists, and
- Communications Equipment Lists.

To create a graphic representation of the unit may require only the fields: name, size, type, location, orientation, velocity, and activity. In most cases, the unit size and type will form an index into a table of standard military symbols. These are placed according to the location field (which is actually short-hand for two or three coordinates). The symbol can then be annotated with the unit's name and iconic representations of velocity and orientation.

The graphics computer is actually a simulation of the visual representation of a military unit. Therefore, this is a crude, but efficient, vertical integration between a high fidelity simulation (ALBAM) and lower fidelity simulation (the graphic computer).

The same is true of the connection between the simulation model and the after action analysis (AAA) computer. In this case, the AAA may require quite a bit more of the characteristics of the simulated unit, but the method for acquiring and transferring this information may be exactly the same. In fact, the AAA system may add detail that is experienced, but not saved or organized in the main simulation. AAA systems are typically concerned with the performance of organizations and units through time. Since the AAA maintains a record of all collected events and statuses, it is in a position to know and synthesize data in the time domain. The driving simulation does not maintain this information. It is only aware of the state it is in right now, and has no memory of past events, though it does reflect the results of having experienced them. The simulated unit's strength may be down to 50%. The unit's data record contains no information as to how it arrived at this state, or even whether this state is its natural condition. Only the AAA system maintains these records and can demonstrate the events that led up to the present state.

We are not asserting that the main simulation **could not** have stored this information itself, only that it **does not**. The same situation exists when interfacing any two simulations. There is no reason that a constructive ground model could not maintain and track the locations of all of the equipment that has been "constructed" into a single unit record with a single location. But, the fact is that this type of information is not essential to the mission for which the model was built. Therefore, these characteristics are handled by another model which does have such a mission. This separation of purpose is one of the driving factors behind the need for vertical simulation integration.

2.3.2 Wide Area Networks

The Corps Battle Simulation (CBS) is used world-wide to create simulated training environments with which to improve the performance of military forces. Like ALBAM, it attempts to model all aspects of the battlefield at an acceptable level of detail. But, it has extended the concept of distribution much further. Though the simulation model may be running on a

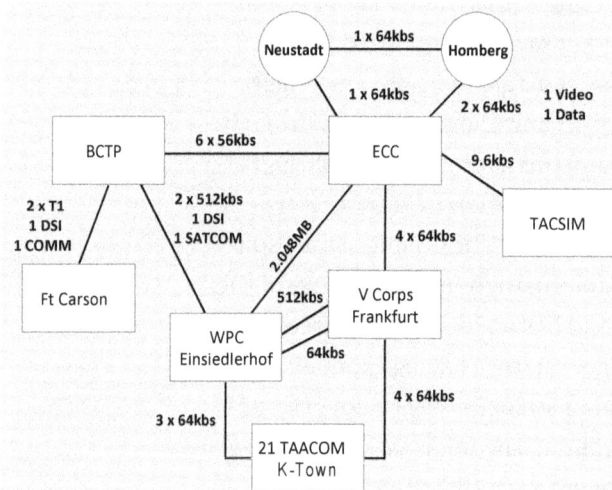

Figure 2-5. Sample Simulation Exercise WAN

computer in Germany, users may interact with the model from terminals and computers in America, Korea, or France. (Figure 2-5). These players are usually equipped with computers which are responsible for providing a local copy of all data needed at that site.

The CBS core computer maintains the master copy of the database. This is then distributed to a process called the "Master Interface" (MI), which is responsible for brokering all data between the core database and the remote databases. When the simulation moves or destroys units, it updates its own database and informs the MI of the changes. The MI then updates its primary database and begins the process of distributing the change to all of the distributed systems that it knows are interested. When the cycle is complete, all computers on the local and wide area network contain their version of the data change first created by the CBS simulation model. This does not imply that all of the databases are identical. The MI understands exactly which remote systems are interested in which types of data and distributes only that information to the site. This reduces the amount of traffic on the networks, and reduces the filtering that must be done by the remote systems to locate the data of interest [CBS Executive Overview, 1993].

To transmit this information, the MI creates data packets in a specified format. Upon receipt, these are then translated by the receiving system and the data stored or processed according to the local software's instructions. A sample of one of these packets is shown in Figure 2-6. This is a common representation used for data packets in networking texts. It begins by identifying the type of data that will be found in the packet; followed by the intended receiver and sender of the information; the body which contains the actual data to be shared; and closes with a trailer indicating that the end of the packet has been reached. We point this out, because this same concept is being heralded as a great advantage of the new Distributed Interactive Simulation (DIS) protocols, which we will examine a little later.

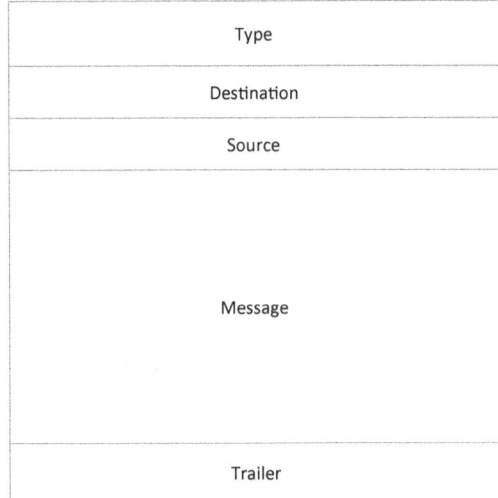

| Type |
| Destination |
| Source |
| Message |
| Trailer |

Figure 2-6. Simulation Data Packet

The organizational structure between CBS, the MI, and the distributed operations stations can be drawn as in Figure 2-7. This type of organization is very effective, in fact it is being imitated in current experiments to connect the constructive BBS and the virtual SIMNET simulations. One beauty of this system is that all interactive and scheduled data requests are handled by the MI. This leaves the CBS model free to spend the lion's share of its time processing simulation events. It's entire connection to the outside world is the MI. Servicing only one customer and accepting no interactive data requests eliminates many possibilities for slowing the system.

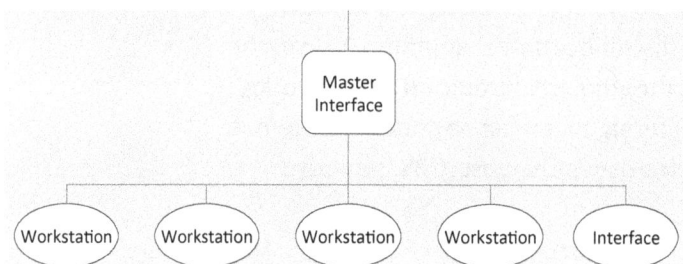

Figure 2-7. CBS Workstation Connections

The Tactical Intelligence Simulation (TACSIM) is attached to several remote components. One of these, the TACSIM Analysis Operations Node (TALON), receives all reports produced by TACSIM; analyzes them; and produces a summary report which is then returned to TACSIM for dissemination. TALON receives the output of TACSIM, making it similar to the graphics terminal, which is only a receiver. TALON also returns reports to TACSIM for delivery to the users. This extends the system to a two-way flow of information. TALON uses the game time provided by TACSIM to insure that its operations are synchronized with those of the core simulation. In fact this game-time is often not determined by TACSIM, but is passed from CBS to TACSIM. Under this configuration, time flows forward, and it flows physically (as data packets) from simulation to simulation. The interfaces create a chain in which each simulation is one time step behind the link in front of it [TALON, 1994].

Model:	CBS	->	TACSIM	->	TALON
Time:	t		$t-\delta t$		$t-2^*\delta t$

where, δt = size of the time step.

This configuration does not disrupt the operations of these aggregate simulations because:

- Each operates independently of the others,
- The confederation is stimulating humans thinking in terms of hours,
- Computer processing speeds prevent lag behind the set time differences.

If we were connecting CBS or TACSIM to a virtual-level simulation, much of the same reasoning could apply. But, if this chaining effect were to be extended through 100 or 500 simulators, it would no longer be acceptable. A virtual simulator's position in the chain would dictate the time it experienced, and another object, physically next to it, may be experiencing events as much as 100 time steps in the future. This interface technique does not scale.

The above interface technique has been used successfully for years to connect 2 to 6 constructive simulations. The time lags at these levels are insignificant when compared to the mission of training combat commanders. These commanders are used to having their situation maps and report boards updated hourly. Therefore, the effect of a six minute lag between two simulators is lost in the noise.

We mentioned that TALON analyzes TACSIM reports and produces a summary. This is a form of fidelity conversion. Details in TACSIM data are condensed into the significant events which are of interest to participants at a higher level of aggregation. TALON adds a level of aggregation and allows the human participants to eliminate the people that formerly performed those operations.

The simulation must retain the accuracy of the TACSIM data input, but must reduce the volume by a factor of 100. Since TACSIM is an intelligence collection model, the data given to TALON represents one side's perception of the enemy. To aid TALON in analyzing this information, it is given access to the true information about the units being detected. With both pieces of data it can then appear to make deductions as its human counterpart would, but without having to resort to the field of artificial intelligence. is more akin to an expert system—given certain inputs, what would humans have decided about the data.

The processing done by TALON is similar to that which must be done when data is passed from a virtual level simulation to a constructive simulation. Many of the details need to be washed out without losing the thread that is of interest in the aggregate world. To do this, the items of aggregate interest must first be identified. Then a mapping can be done from all object data to these interest items, with some information being ignored. As an example, TALON summaries consist of the following data items:

- UnitName,
- Location,
- Orientation,
- Activity,
- Affiliation,
- Time of State,
- Size, and
- Type.

TACSIM, on the other hand, produces much more detailed information: the radar characteristics of specific objects on the battlefield; the arrangement of individual pieces of equipment; and the command structure between several units. Filtering this down to the TALON level requires the omission of details.

The TACSIM-TALON interface is a special case of data flowing from a virtual level simulation to a constructive simulation. We state this even though TACSIM and TALON are both considered constructive simulations. Actually, TACSIM just operates at a lesser degree of aggregation than TALON. Dividing simulations into the three categories—Constructive (Aggregate), Virtual (Object), and Live—creates a simple frame of reference. But, it also categorizes simulations more strictly than is naturally true.

Connecting TALON to TACSIM is done via a networking technique known as remote procedure calls (RPC). These allow an application to be written as if all of its components are operating on the local machine, though they actually span a network. The communication utilities are hidden from the system developer and user. The structure of RPC is shown in Figure 2-8. This technique is extremely useful when a distributed simulation requires that you communicate with remote components as easily as with local [Shirley, 1994].

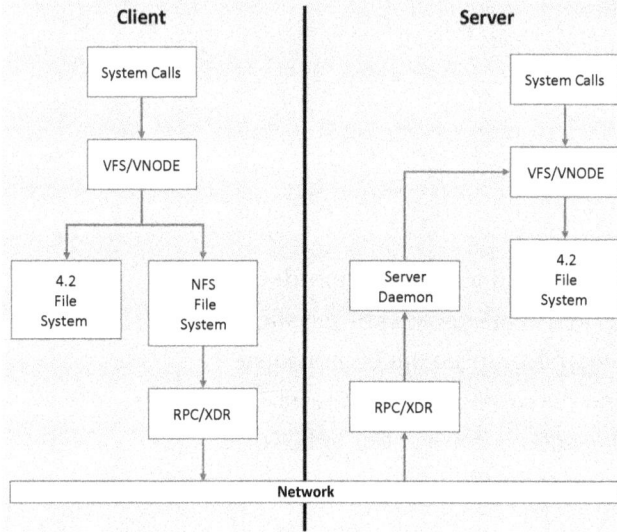

Figure 2-8. RPC Structure

RPC is part of a larger technique known as client/server programming. A server is assigned the task of processing requests for information and sending the results to the requester. The requester is known as a client because it comes to the server asking for assistance, much like going to a lawyer or accountant for services. The client requests, the server complies. A server may receive requests from any number of clients on any number of network nodes.

On the ever-popular Internet, most of the services are established using this technique. The ability to access another computer and download files to a local computer is known as "ftp" (file transfer protocol). It appears to the user that he has accessed the remote computer and is exploring its contents. Actually each command is being brokered between the local ftp client and the remote ftp server. Each command entered is interpreted and transmitted to the remote computer. The appropriate response is then transmitted back to the local client where it is displayed, providing the appearance of being "logged in" to the remote computer. The login process was simply a security check to insure that the client had permission to access the server application. Other client/server applications include: finger, telnet, gopher, WAIS, WWW, and archie. When simulations become distributed widely these same techniques must be used to verify permissions, provide timely responses, and hide the distinction between remote and local services.

2.4 Dedicated Horizontal Interfaces

Although interfacing all simulations is a new effort in DoD, limited interfaces have been built in the past. These have usually consisted of software and networks developed jointly by two model proponents to meet their interfacing needs. These interfaces operate only between the two target simulations, and are not able to connect others into a loose confederation. Unique interfaces exist between CBS and TACSIM, CBS and AWSIM, CBS and CSSTSS, CBS and RESA, TACSIM and GRWSIM, and others. Although the software can not be reused, the techniques and lessons learned are valuable.

2.4.1 CBS—AWSIM Interface

With CBS handling the ground warfare and AWSIM handling the air warfare during an exercise, the interface between them is an excellent example of the type of two-way communication and sharing of control that can be useful in a generic vertical integration. One of the basic concepts used in this interface is the idea of "ghosting". A ghost is an entity in one simulation that is in no way controlled by that simulation. It may appear or disappear, without any cause, from the local simulation. An excellent example of a ghost unit is a flight of fighter aircraft being flown and controlled by AWSIM. The air model determines when and what type of aircraft will be launched from an air base. Once airborne, these are controlled by pre-loaded or interactive commands directing them on their mission. Once the mission is complete, these aircraft return to the air base and lose their unique identity as they return to the inventory of the air base.

While airborne, these aircraft are subject to attack by enemy aircraft and by ground-based air defense systems, including surface-to-air missiles (SAMs) and anti-aircraft artillery (AAA). SAMs are modeled in AWSIM, and engage the aircraft as a natural component of the model. But, AAA assets are usually controlled by ground units only in CBS. If the aircraft do not appear in CBS, there is no way for the AAA to impact the outcome of the war [ALSP, 1993].

To create a target for the AAA, the AWSIM aircraft were ghosted in the CBS model. They appear from nowhere upon launch and disappear upon landing. During their mission, the ghost may be detected and fired on by the CBS AAA assets. But, since the aircraft are ghosts they can not be hit in CBS. Instead the engagement event is sent to AWSIM to determine the outcome. In a sense, the AAA fire is then a ghost in AWSIM, appearing from nowhere and causing events. When AWSIM receives the AAA event message, it uses its native algorithms and response surfaces to determine which, if any, aircraft are killed. Those that are killed are then removed from the AWSIM model, just as if they had been killed by a local SAM battery. A message is then generated and sent to CBS so the same results will be reflected in the ghost.

In a typical battle, many different interactions can occur
between CBS-controlled and AWSIM-controlled entities.
Each of these is handled in a similar way, allowing interaction between models, but maintaining their basic independence. The table below illustrates the ghosting that
occurs to allow these interactions [CBS-AWSIM, 1993].

CBS	AWSIM
Ghost	Fixed Wing Aircraft
Rotary Wing Aircraft	Ghost
SHORAD	None
HIMAD	HIMAD
ALLRAD	ALLRAD
Acquisition Radar	Ghost
Ghost	Airlift
None	Air Bases

With this list, we define the points at which interaction
can take place between the two models. Fixed Wing
Aircraft and SHORAD (Short Range Air Defense) were
described in the example. There is no need to ghost
SHORAD into AWSIM because the targets they are looking for are brought into CBS for them. This same idea
defines the interface for helicopters and acquisition radar.

AWSIM air bases are not represented in CBS. But, in real
world conflict helicopters may land at air bases to receive
fuel and weapons. To enable this capability, CBS has
agreed to place a Forward Area Rearming and Refueling
Point (FARRP) at each air base location. Both items are
fixed and thus simulate the use of air bases by helicopters.
The modeling of HIMAD (High and Medium Range Air
Defense) assets is a combination of the ghosting and joint
location techniques. The asset actually appears and can
be controlled in both simulations. But it has different

operational characteristics in each. To AWSIM, HIMAD assets are seen as missile emplacements which can be fired, replenished, turned on and off, and be destroyed. To CBS, HIMAD assets appear as ground vehicles which can move and be destroyed. In this sense, CBS owns the wheels and drivers, while AWSIM owns the missiles and fire-controllers. When CBS moves its portion, the AWSIM portion moves with it. When AWSIM fires missiles, CBS loses these supplies. When the AWSIM HIMAD is bombed, both the top and bottom of the asset are destroyed. When the CBS asset is destroyed by artillery fire, both pieces are destroyed. When the HIMAD runs out of missiles, these are resupplied by CBS. A message passed to AWSIM then replenishes them in that model where they are actually used.

The HIMAD assets seem to be more fully integrated than other assets. This is not true. The nature of the asset calls for a two-part rendering, where the nature of the other assets does not require this level of interaction.

The ALLRAD (All Range Air Defense) assets are modeled in CBS as SHORAD, and in AWSIM as HIMAD. They are then toggled between looking for targets in CBS or in AWSIM. The theory is that, although capable of shooting at all targets, each is actually assigned a mission which dictates where it will operate. Changing this mission transfers control of the asset from one model to the other. The CBS model may mix air defense asset types that are contained within a unit. AWSIM allows only one type of air defense asset per unit. To join the two, CBS agreed to place only one type of missile in each unit joined with AWSIM. This type of limitation is indicative of the problems that can occur as we explore a generic interface between many models.

2.4.2 CBS—CSSTSS Interface

The interface between CBS and CSSTSS resembles that
between CBS and AWSIM, in the techniques employed.
The Combat Support Services Training Simulation System
(CSSTSS) is a very high fidelity model of the mission
performed by logistics assets. It is responsible for resupply
and maintenance missions. When we mentioned above
that CBS resupplies the missiles used by AWSIM, this assumed that CSSTSS was not also operating in the exercise.
If it were, the mission would actually be controlled by
CSSTSS, but the results would be the same from AWSIM's
point of view [CBS-CSSTSS, 1993].

When CBS realizes that a unit needs to be resupplied,
it sends a request to the supply unit assigned to it. This
message, containing a list of the needed equipment, food,
fuel, etc., is then passed on to CSSTSS which is responsible
for acting on it. CSSTSS prepares the logistics package and
loads it on the delivery trucks. These trucks are subject to
repair and maintenance schedules in CSSTSS. Therefore,
although they exist, a delay may occur until they are ready
for operations. Once CSSTSS has determined that the assets can be placed in a convoy on their way to the combat
unit, a message is sent to CBS which generates the supply
unit. CSSTSS determines the route that will be taken, the
unit that will be serviced, the number and type of trucks
that will be used, and the time to load and unload the
supplies. Once instantiated in CBS, this convoy of trucks
is subject to its movement constraints and enemy assets.
To CBS, the unit is just a convoy of trucks. The contents of
these are only represented in CSSTSS.

If the convoy is attacked and destroyed, CBS relays the

information to CSSTSS which then determines exactly which supplies were destroyed. When this convoy finally arrives at the combat unit, it will contain a reduced amount of the requested materials. CSSTSS will then determine when these are actually unloaded and available for use by the CBS unit.

Should the convoy be obstructed such that delivery of the supplies is impossible, both models agree that the convoy will retrace its steps and return to the supply unit that generated it. This prevents missions from hanging in the simulations with no clear success or failure.

CSSTSS also controls the helicopters that are used to airlift combat units in CBS. In a sense, the CBS unit is then viewed as supplies for delivery. CSSTSS determines whether the helicopters are available, based upon their flight histories and maintenance schedules.

2.4.3 CBS—TACSIM Interface

CBS and TACSIM have a one-way interface in which data flows from CBS to TACSIM. Before an exercise the databases are built jointly such that TACSIM's contains all of the units in CBS's. During the exercise CBS sends location changes and attrition results to TACSIM, which mirrors them in the corresponding unit [CBS-TACSIM, 1993]. The one-way interface is required for security reasons. This also dictates a very simple architecture compared to the two examples above. Message passing and control are simple tasks here. But, there is another aspect of this interface which is of interest to this work. CBS tracks a unit as a "bucket" which contains an assortment of objects. Although these objects are individually identified, they can not be individually controlled. But, each object has attributes which allow it to contribute to events such as unit combat and movement.

TACSIM tracks the unit as a single controllable piece, also. But, it also tracks the location of each individual object. It does this by creating a template into which the CBS objects must fit, according to their activity. A unit moving in a convoy would be assigned a template that roughly forms a line in which each object is oriented in the same direction, single file nose to tail. The same unit, in attack posture, may be arranged by a template into a phalanx. This allows TACSIM to collect information which intelligence analysts can use to deduce these unit characteristics. The templates are a way of adding detail to a CBS unit-bucket.

CBS also identifies each unit-bucket with a single 16 character name. This is fine for tracking entities within a simulation, but is not adequate for intelligence analysts, who are being trained to deduce unit identifications and relationships based on this information. Therefore, TACSIM adds detail in this area, mapping more descriptive identifiers to each CBS name and adding several levels of command structure within the unit-bucket which CBS does not replicate at all. This addition of detail may be useful in connecting virtual and constructive simulations as well.

2.5 Parallel Simulation Techniques

Multiple, distributed, networked simulations are a form of parallel computing. Having realized the limitations of a single machine, separate models were developed which fit well within the capabilities of the computers available. Joining these into a single simulated environment is a specialized form of a parallel computer. In fact, this field declares one of its sub-domains to be that of distributed computing. Various languages such as p4, pvm, Posybl, and Linda have been developed to facilitate work in this area. These, and other tools, may be very useful in developing distributed simulation interface techniques.

Distributed computing is an effort to exceed the limitations imposed by single CPU computers, but without resorting to the cost of specialized, multiple processor machines. As networking becomes more refined, the idea of harnessing all of the unused CPU cycles on the networked machines is a target of opportunity. One subset of the problems addressed in this field is known as parallel simulation. Here researchers attempt to partition a simulation into logical processes (LP's), which can be assigned to separate processors. These then work on multiple pieces of the problem in parallel, decreasing the time to finish the computations, or increasing the size of the problem that can be addressed within the time constraints [Reynolds, 1993].

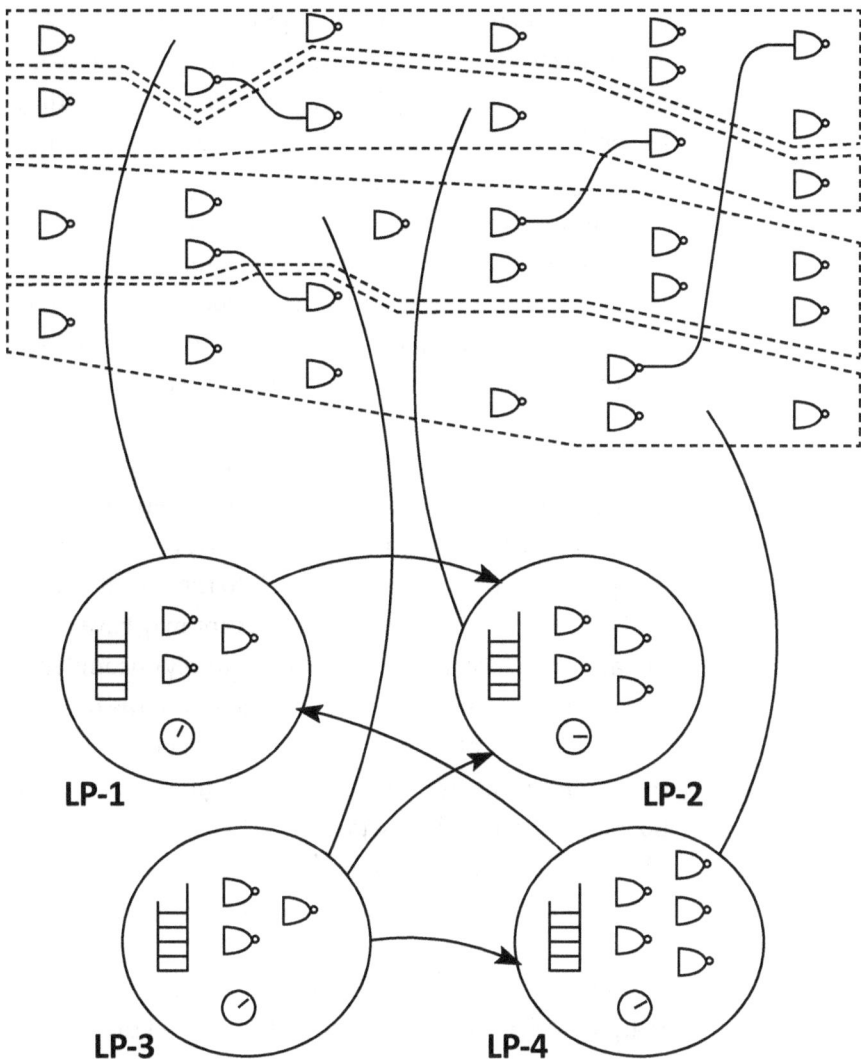

Figure 2-9. Logical Processes in Parallel Simulation

If LP's can be defined such that no communication or data sharing is required between them during the assigned computations, the simulation is said to be trivially parallel. This situation is ideal, and as, with most ideals, rare. Usually, some degree of communication is necessary to provide the data needed for each process to continue. At the very least, knowledge of the time being simulated on other CPU's is necessary. The typical approach to dividing the problem into LP's is to look for natural spatial separations (Figure 2-9). In military combat, the domain of an event is limited to some geographic area. The degree to which spatial decomposition can create sets which do not interact is the degree to which independence is achieved between LP's. Many different types of LP division will be explored as one of the primary topics of this work.

One controlling factor for LP's is the fact that eventually events or entities will cross the boundaries between them. When this happens it is necessary that the time stamp associated with the event or entity fit into the time domain of the new LP. If this is not true, the event may have no impact, or an incorrect impact, on the newly encountered LP. Assume that an artillery salvo is filed from the domain of LP-1 into that of LP-2. If LP-2 is operating many game-time hours ahead of LP-1, the shells will impact in an area no longer occupied by the target. In this case, LP-1's view of the world at time t must be able to generate effective events in LP-2 which will also occur at time "t".

To maintain this time synchronicity, ideas such as Time Warp were developed. Here, each LP is allowed to calculate events according to the power of its CPU and constraints placed on it by the environment encompassing the entire simulation. All processes are constructed such that they periodically exchange their local time. These times are then used to determine the Global Virtual Time (GVT). We will not explore all of the steps in calculating an accurate GVT here, although some will be detailed in exploring new techniques for vertical integration [Jefferson, 1985].

LP's may proceed conservatively or optimistically, once GVT is established. Conservative progression limits each LP to operating in a time window around GVT. Once the threshold of this window is reached, the LP must put itself in a "wait" state until GVT progresses. This insures that all LP's will be operating within the same limited range of time, and that boundary crossing events can be implemented in the LP's receiving them.

Optimistic progression assumes that boundary crossing events are rare, and therefore, each LP should use its processing power to move as fast and as far along the time axis as it can. Should one of the rare events arrive, the LP will then backtrack to a time just before the event and reprocess everything after that, with the new event included. Obviously this rollback process can be very computation and storage intensive, making optimistic time warp something to be balanced against the frequency of boundary crossing events.

One of the requirements of distributed interfaces is to maintain the necessary level of synchronization between all of the remote pieces, this is also one of the major topic in the parallel computing field. Distributed software libraries, such as *p4* and *pvm*, send messages to specified processes, providing them with important information, such as the time and events that they must handle. An example of each of these library calls is:

```
int info = pvm_send(int tid, int msgtag);
int info = pvm_recv(int tid, int msgtag);
int info = pvm_sendsig(int tid, int signum);

p4_send(int type, int to, char *msg, int len);
```

```
p4_recv(int *req_type, int *req_from, char **msg, int
*len_rcvd);
```

These require the following variables,

tid, to, from = Destination for Message,
msgtag, signum, msg = Message Pointer,
len, len_rcvd = Length of Message, and
req_type = Message Type.

This is as simple as a standard function call in C programs, but the data passed actually arrives at a process running on a different computer. Both p4 and pvm hide the network communications from the programmer [Butler, 1992 and Geist, 1993].

There is one limitation to man-in-the-loop (MIL) simulation which is usually not a factor in parallel computing and parallel simulation. MIL applications are expected to run in real-time, where real-time is defined as simulation time passing at the same rate as clock-time, and using a time step that is appropriate for the level of aggregation of the simulation. If the computers are capable of performing computations much faster than this, all of their power can not be used.

Extra power beyond that needed to maintain real-time operations can be harnessed to some degree. If some form of optimistic time warp is being used, the CPU can calculate events in the future. This information must then be queued up for later release to the MIL, since it depicts his future, which he is not allowed to know. But, should events from another LP cause a rollback, this pre-calculation must be "un-done" by returning to the state before

the new event's time-stamp. This may endanger the ability
of the processor to maintain real-time synchronization for
one or more time-stamps. Since the pre-calculated events
were not released to the MIL, he will never know that
these were preformed, unless the process subsequently
falls behind real-time.

In the vertical integration environment the major components are DIS and ALSP. Simulations on DIS are expected
to maintain accurate event data with update time intervals
of 100 or 300 milliseconds. ALSP simulations, on the other
hand, maintain a time step of 1 to 6 minutes. Obviously,
ALSP provides a time buffer, which may allow rollback
and recalculation to be performed without impacting the
rate of time flow, DIS may be at risk. This may indicate
applications for optimistic time warp in messages flowing
to ALSP, but not for those flowing to DIS.

When constant real-time is required, the power of the
computer selected will be impacted by the type of task
it will perform and the type of software algorithms to be
used.

Given that we are using a conservative approach, we must
acquire a computer that can handle the maximum load
during any given time step. With optimistic approaches,
we may be free to acquire a machine which provides
real-time operations for some average load over some
number of time steps which can be integrated.

Again, once the computers are selected, we have limited
the size of the scenario and the level of fidelity which can
be represented. Exceeding this size limit will require more
LP's distributed over more CPU's.

On method of parallel simulation is patterned after the Single Instruction Multiple Device (SIMD) computer. Here, one event stream is managed by a single processor. This then serves the events that they are allowed to process to all of the other client processors. Here, it is impossible to use optimistic time warp techniques, since each CPU is knowledgeable only of the events that the event manager considers "safe" to work on. In many ways, this is identical to multiple simulation processes operating on the same CPU. The only difference is that access/modification to the database is being managed remotely.

In any case, parallel computers are faced with situations in which no two LP's are glued to a universal time or rate of time progression. Each may be free to live as fast as it can. But when these LP's interact, they experience some of the same confusions that Kurt Vonnegut's Billy Pilgrim did in *Slaughter House Five*. He was continuously coming "unstuck in time", losing all sense of causality and responsibility to the events occurring around him. In T.H. White's, *The Once and Future King*, Merlyn the Magician lives his life backward in time. Beginning life as an old man, and dying as a newborn baby at some later date. As his life progresses, he is continually forgetting people and events that happened in the conventional past, which is his future. When simulations experience this, there can be no guarantee of causality and we cease to replicate reality as we know it.

2.6 Aggregate Level Simulation Protocol

2.6.1 Origins

In the late 1980's, DARPA conducted some very successful experiments in linking armor simulators within a
single virtual environment, a project known as Simulator
Networking (SIMNET) project. As a result, they ap-
proached MITRE with a request for a standard interfacing capability between constructive simulations. models
which they expected to integrate were primarily training
tools used to stimulate soldiers in their various positions in combat. Training exercises, such as Reforger and
Ulchi Focus Lens, were typically carried out by one or two
large simulations, such as CBS and AWSIM. This project
intended to join any model with the correct interface into
the exercise. The project is now known as Aggregate Level
Simulation Protocol (ALSP) [Weatherly, 1991].

ALSP is based on four fundamental design principles
gleaned from the SIMNET work:

1) Distributed Computation Based on Entity
   Ownership,
2) Avoidance of Single Critical Resources,
3) Reliance on Broadcast Communications, and
4) Replication of a Limited Set of Attributes Among
   all Simulations.

These principles are intended to overcome many of the
limitations seen in earlier efforts to join simulations. The
first, implies that joining the ALSP confederation will not
significantly increase the computational burden of the simulation. The second, eliminates reliance on resources which

can become bottlenecks to the entire system. The third, is intended to allow flexibility such that a simulation may join or leave the confederation during an ongoing exercise. The fourth, is intended to create a common foundation among all simulations for a limited set of entity attributes.

2.6.2 Operations

The operation of ALSP is divided into three categories.

Data Management controls which simulation is allowed to modify which data fields for an entity. Each simulation divides entity attributes into two categories, public and private. Private attributes are those known only to the local simulation and not shared with the network. Public attributes are for the common foundation mentioned earlier. For example, the following table may indicate which, of many attributes of a unit, are maintained by a simulation:

CBS	TACSIM	AWSIM	CSSTSS
ID	X	X	X
Location	X	X	X
Size	X	X	X
Type	X	X	X
Strength	X	X	-
Orders -	-	-	
Orientation	X	X	-
-	-	Altitude	-
-	-	Weapons Status	-
-	Name	-	-
-	RF Frequency	X	-
-	Deploy Pattern	-	-
-	Down Link	-	-
-	-	-	Inventory
-	-	X	Maintenance

Here "X" indicates that the characteristic is known in some form to the simulation, and "-" indicates that the simulation does not deal with data of this type (Figure 2-10).

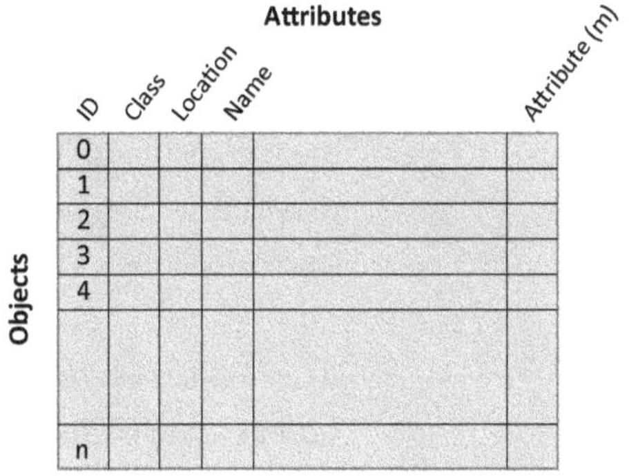

It is the public attributes that require management. Although all simulations may be aware of the existence of the entity and the state of its public attributes, this does not imply that they may modify these attributes. Typically each entity is controlled by one simulation which is then respon-

Figure 2-10. Aggregate Level Simulation Protocol Design

sible for maintaining its attribute states. Events effecting this entity are then routed to its parent simulation for actualization.

Let us remember the CSSTSS convoy generated by a request from CBS, in the section on Dedicated Horizontal Interfaces. There a special message had to be generated to request CSSTSS convoy resupply. If characteristics of this operation effected any other model, a separate message would have been sent to those. But, with the advent of ALSP, all simulations are now on the same network and speak the same protocols. A single standard message can now be generated to activate all necessary events among all confederation members [CBS-CSSTSS, 1993].

Time Management controls the temporal causality of events among all simulations. This function is performed for all messages transmitted across the network. In earlier sections, we described conservative and optimistic methods for handling time in a parallel environment. ALSP chose to use the conservative method. The use of optimistic methods would have required that all simulations joining the network operate state-saving and rollback algorithms. Most of these types of simulations do not have such a capability and dictating the creation of these algorithms to join the network is infeasible. The simulations also operate with men-in-the-loop. Therefore, each simulation would have to create some type of moving window to store events and states until they could be released to the humans.

ALSP manages the time of all simulations to create a Global Virtual Time (GVT), which all simulations are then subject to. To accomplish this, all simulations are assigned a beginning time, assume day 1 hour 1 minute 0 (Julian 0010100). From there, each is allowed to simulate the events taking it forward one time step. But, each simulation is free to use a step size appropriate to its operations. Therefore, the simulations may step forward out of sequence. But, the minimum of the new current times becomes the new GVT, and is distributed to all simulations [Weatherly, 1991].

ALSP Time	t
CBS Ops	$tc \geq t + \delta tc$
TACSIM Ops	$tt \geq t + \delta tt$
AWSIM Ops	$ta \geq t + \delta ta$

Assume that the CBS time step is 6 minutes, AWSIM is 1 minute, and TACSIM is 2 minutes. The new GVT after one step would be 0010101, which would be distributed to all. CBS would realize that is free to calculate from 0010101 to 0010107. But, its 6 minute time step would also dictate that it does not step again until 0010106. Similarly for TACSIM. But, AWSIM would see that it has arrived at its next desirable time step, and calculate again. Therefore, AWSIM would loop through 6 time steps and TACSIM 3, before CBS would move again. But, when CBS did move again it would include the accumulated events from the other models. Using this method each simulation is free to see the world in chunks that are significant to it. This also allows them to operate in the same manner that they were prior to joining the ALSP confederation, resulting in a minimal amount of software changes. It is estimated that less than 10% of the simulations software will be impacted by joining ALSP.

Architecture is the structure used to join all the simulations to each other and to the network. This consists of protocols, translators, and gateways. Two protocols are used: that between the translator and the gateway, and that between gateways.

Each translator is a custom entity which is built to allow existing simulations, such as CBS, TACSIM, and AWSIM, to join the ALSP confederation. The translator changes local data semantics into global semantics understandable by other simulations. It also performs the reverse translations on incoming messages. This operation will be greatly reduced, once simulations are built specifically to join ALSP.

Each gateway transports message packets across the network to all other simulations. It manages the addition and removal of simulations during an exercise. Finally, it manages the local time, and notifies the network manager of this, so that a GVT can be reached.

In ALSP, simulations are known as actors. They need not be traditional simulations interested in controlling objects and interfacing with other simulations. They may be data collectors or graphic workstations which are receive-only systems. The gateways are known as ALSP Common Modules (ACMs). There is also an actor that is actually part of the ALSP architecture. The ALSP Broadcast Emulator (ABE) calculates and distributes the GVT for all participating simulations.

Since objects are owned by a simulation, this dictates some basic limitations on adding and removing simulations from the confederation. Prior to the beginning of an exercise, control of all units must be negotiated and assigned. Once an exercise begins, any simulations joining may not assume control of entities which are already owned by another simulation. They must, therefore, bring with them the entities they will be responsible for. This implies that the addition and removal of simulations must be managed by some exercise controller. Otherwise, entities may appear suddenly in the middle of ongoing interactions, destroying the validity of the operation. Similarly, when a simulation is removed from the confederation, all entities that it controls must be removed with it or control passed to another member of the confederation. This may include transferring control to Computer Generated Forces (CGF) models.

The basic steps in entering the simulation environment are:

1) *Join the Confederation.* This is done by sending a "join" message, which notifies all other simulations and the network of your existence.

2) *Refresh Local Data.* This is a "refresh" message, which causes all simulations to transmit the public

data elements of the entities which they control. This information brings the new simulation up-to-date with the situation as it now exists.

3) *Operate.* The new simulation now operates in the environment as an equal member of the confederation.

4) *Remove.* A simulation exits the exercise by sending a "remove" message to the network so that all packet traffic intended for it will be eliminated. This assumes that the final part of step 3 was to transfer ownership of any controlled entities that need to remain in the exercise.

The ALSP Operational Specification gives the impression that, although intended to be generic and universal in scope, the project has actually evolved into a proprietary solution to a known problem. No explanation is given of the general outline of the message packets to be delivered. The entire document describes library calls for executing a set of known simulation interactions. This implies that ALSP may be becoming a custom interface such as those described above, but involving a dozen simulations rather than just two. Although such a library of routines is necessary regardless of the generic nature of the interface, it is the lack of a general description that leads one to believe that the generic focus has been lost. No attention is given to unknown simulations and undefined interface events.

The Defense Modeling and Simulation Office (DMSO) commissioned a panel to review existing aggregate level linkage technologies. This panel made several pertinent observations. First, all current linkage work addresses the perspective of the local simulation organization. None of them is investigating the larger, more general problem of connecting simulations at all levels. A senior representative of DIS made the statement, *"All we are doing is anecdotal; there is very little theory"*, emphasizing the need for studies of this type [DMSO, 1993].

The panel concluded that the distinction between vertical and horizontal interfaces is artificial. Simulations can not be neatly divided into virtual and constructive categories. Rather, every simulation has its own unique representation of reality, and interfacing any two of them presents unique situations and problems. The recommendation was to develop a common conceptual language for representing combat phenomena, objects, and relationships. Once this is done, simulations can be built which will interoperate naturally, regardless of their level of aggregation. Although there is a problem of aggregation and disaggregation to be solved, this is complicated and confused with the problem of differing conceptual languages between the simulations. The language must be standardized first.

An example of a problem that is both constructive and virtual is found in the air transportation field. An aircraft in flight is seen by those aboard to be a group of hundreds of individuals and aircraft sub-systems. But from the air traffic control tower it is a single entity with a very limited set of attributes, few of which are replicated inside the aircraft. By the same token the airport appears as a single entity to the aircraft through most of its flight. But, near landing it suddenly becomes a set of runways, competing aircraft, and control personnel. Is this a constructive or a virtual problem? Which direction is vertical and which horizontal?

An essential characteristic of all simulations must be that a problem may be solved with any combination of simulations that are appropriate. The solution provided by any of them should tend toward the same direction. Although, the outcome of specific events may not be the same, the global outcome should be. This is certainly not true with the simulations available today. There are no standards which govern the replication of reality within a simulation. Therefore, the variability that exists prevents any sort of similarity between simulation results. Whether one chooses a suite of: CBS+AWSIM, CBS+RESA, BBS+AWSIM, CBS+TACSIM, CBS+BICM, etc., the results should vary because of the Monte Carlo nature of the simulations and the input provided by the human operators. They should not vary because of the different methods of representing reality.

The DMSO study group concluded that in establishing a common foundation of entity attributes, several characteristics should be considered:

1) Minimal software changes to existing models,
2) Maximum software reuse,
3) Open systems approach,
4) Distribute globally,
5) Scalable,
6) Extensible to, as yet, unknown simulations.

2.7 Distributed Interactive Simulation

Undeniably, the largest simulation interfacing effort,
Distributed Interactive Simulation (DIS), is focused on
joining virtual level simulators. These include: armored
vehicles, helicopters, aircraft, naval vessels, and individual
humans. The project was initially sponsored by DARPA,
and has since transitioned to the Defense Modeling and
Simulation Office (DMSO) and the Army Simulation
Training and Instrumentation Command (STRICOM).
The architecture, protocols, and representation standards
are being developed by industry, government, and university representatives who meet at semi-annual workshops
to exchange ideas. These workshops started with 55
participants in 1990, and have grown to over 1000 in 1993.
Most of the work is being funded by simulation project offices, who intend to become members of the DIS network
when it is operational, and by industry investment, in
hopes of winning contracts in the future [The DIS Vision,
1993].

DIS strives to create a cohesive environment in which soldiers can train against an enemy, using any type of equipment and tactics. The military hopes to present them with
realistic combat events, which can not be reproduced in
a physical training environment because of safety, equipment, terrain, and budgetary constraints. It will become
an environment in which joint and combined exercises
can be conducted. joint exercise is one in which multiple
services participate—the army, air force, navy, marines,
national guard, and coast guard. A combined exercise
includes U.S. allies, particularly members of NATO and
South Korea.

There are three primary advantages to performing virtual exercises rather than physical ones. First, they are relatively inexpensive. The military watched the cost of the large, German exercise Reforger drop from $115 million to $22 million, as they added more simulations to replace physical activities. With DIS they hope to train pilots in a simulator, and experience this same level of savings. Second, the exercise can be distributed. For a physical exercise, all of the participants must be brought together on a common playing field. Transporting and housing the troops and equipment is a major undertaking, requiring long planning periods and large pieces of terrain. Networked simulators can exist anyplace in the world, and yet appear to be operating together in the same area. This need is increasing as the NATO countries no longer perceive a serious threat from the Soviet block. Their farmlands had been a primary playing field for these very destructive exercises. Third, simulations are more secure than physical exercises. When practicing combat tactics across large physical areas our opponents usually send observers who collect this information and use it to plan their defenses and counter moves. If events occur inside a family of networked computers they can be masked by limited network access and encryption.

In addition to the training benefits, the Army Science Board sees this type of simulation as useful in the system acquisition process, the test and evaluation of new equipment, and combat requirements development.

DIS is founded upon six basic concepts.

1) No central computer for event scheduling or conflict resolution,

2) Autonomous simulation nodes responsible for maintaining the state of one or more simulation entities,
3) A standard protocol for communicating "ground truth" data,
4) Receiving nodes are responsible for determining what is perceived,
5) Simulation nodes communicate only changes in their state, and
6) Dead reckoning is used to reduce communications processing.

Supplementing these concepts are the eight key design principles.

1) Object oriented entity design,
2) Defined sphere of interaction for entity cause and effect,
3) Common beginning gaming area environment,
4) Database driven model designs,
5) Synchronous and asynchronous interconnections,
6) Local models determine level of resolution and aggregation,
7) Simulation management for cross-entity functions, and
8) Communications services will provide functionality required for interconnection based upon the OSI network model.

All of these lead to a DIS architecture as visualized in Figure 2-11.

The DIS network will function like the Internet in many ways. The intent of both is to support users world-wide without allowing local node problems to disrupt the operations of the entire network. A DIS network is designed to support multiple independent exercises simultaneously. To do this, each exercise is assigned an "exercise number". To use the

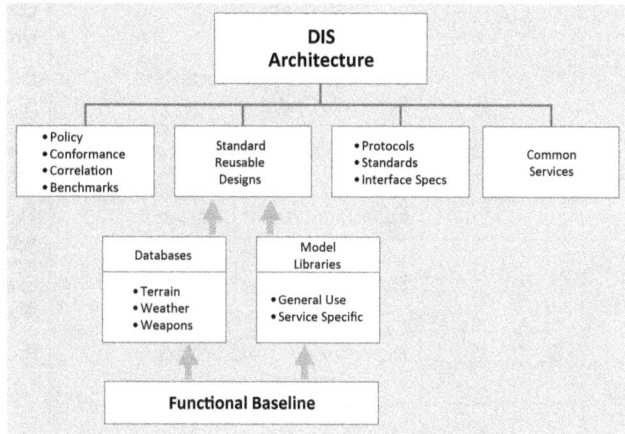

Figure 2-11. DIS Architecture

network, an exercise sponsor must submit information which will allow the network controllers to determine the bandwidth that will be required, based upon the number and type of nodes, the geographic area to be covered, and the level of aggregation. If sufficient bandwidth is available during the period requested, the exercise is approved, if not, negotiations will determine the best alternate time for both parties.

In order to participate in a DIS exercise a minimal level of interface functionality has been defined. These are enumerated in Figure 2-13 and 2-14. The functionality is designed to insure that no node takes advantage of others by rendering some lower level of entity interaction, making it impossible to detect, kill, or communicate with.

FUNCTION	REQUIREMENTS
Interface with Other DIS Simulations	Operate in Real Time Use Standard Comm. Protocol Use Standard Protocol Data Units (PDUs)
Determine Target Location Between Updates	Execute Dead Reckoning Algorithms
Determine Hit or Miss	Execute Weapons Flyout Models
Calculate Impact Damage	Execute Battle Damage Assessment Models
Detect Collisions	Execute Collision Detection Algorithms
Determine Terrain Effects on • Weapons Flyout • Emissions Propagation • LOS Intervisibility	Process Terrain Model
Determine Atmosphere Effects on • Emissions Propagation • Visibility	Process Atmosphere Model
Determine Ocean Effects on • Emissions Propagation • Background Noise	Process Ocean Model
Display to Live Participants • Visual Appearance of Entities	Process Ocean Model Render Visual Image
Atmospheric Effects on Vis	Process Atmosphere Model Process Sensor Model Render Visual Image
Terrain and Features	Process Terrain Data Base Process Sensor Model Render Visual Image
Sea State Effects on Detection	Process Ocean Model Process Sensor Model Render Visual Image

Figure 2-12. DIS Functional Requirements

FUNCTION	CREATE/ TRANSMIT PDUs	RECEIVE/ PROCESS PDUs
Entity interactions Appear on Other Display	Entity State	
Display Other Entities		Entity State
Fire at Other Entities	Fire	
Display Firing Flash		Fire
Damage Other Entities	Detonation	
Conduct BDA* on Self		Detonation
Notify Others of Emissions	Emission	
Sense Emissions of Others		Emission
Notify Others of Radio Trans	Transmitter	
Sense Radio Trans of Others		Transmitter
Send Radio Messages Over DIS	Signal	
Receive Radio Messages Over DIS		Signal
Communicate Receiver State	Receiver	
Receive Receiver State		Receiver
Notify Others of Laser Emissions	Laser	
Sense Laser Emissions		Laser
Notify Others About Collision	Collision	
Determine Collision Damage		Collision
logistics functions		
Request Logistics Support	Service Request	
Sense Logistics Request		Service Request
Provide Resupply	Resupply Offer	
Receive Resupply		Resupply Offer
Indicate Supply Received	Resupply Received	
Understand Supply Received		Resupply Received
End Resupply Action	Resupply Cancel	Resupply Cancel
End Repair Action by Receiver	Stop Sending Service Req.	
End Repair Action by Supplier	Repair Complete	
Understand Repair Complete		Repair Complete
Indicate Repair Result	Repair Response	
Understand Repair Result		Repair Response

Figure 2-13. DIS PDU Requirements

Communication between simulation nodes is done with a set of standard Protocol Data Units (PDUs). These are designed to provide a foundational view of the battlefield for all participants. Each node is free to inject any detail that is only needed locally, as long as it does not distort the public view portrayed by the PDUs. The most commonly used of these is the Entity State PDU. This describes the object of each node to all other nodes. Some of the other PDUs are listed in the Figures above, but there are others which are designed for very specific needs. Since DIS is still in the design phase, PDUs are changed and added at almost every workshop.

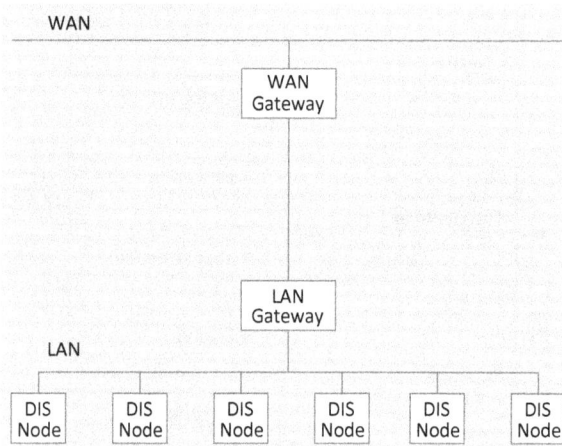

Figure 2-14. Physical DIS Network Interface

As with all communication, latency in data transfer can be a problem. DIS establishes two levels of interaction with guaranteed minimal latency of 100 and 300 milliseconds, and a low variance in these numbers. Entities that are considered "tightly coupled" must maintain a 100 millisecond communication rate. These may include fighter aircraft flying in formation. "Loosely coupled" entities are allowed to communicate at a rate of 300 milliseconds. These are entities in the same sphere of interaction but who are not interacting with each other, such as a tank and a fighter aircraft. The latency variance has not been qualitatively defined at this point.

Simulators perceived to be able to operate in this environment in the near future include the Close Combat Tactical Trainer (CCTT), Mobile Automated Instrumentation System (MAIS), Tactical Combat Training System, (TCTS), Battle Force Tactical Trainer (BFTT), WARBREAKER, and Special Operations Forces Aircrew Training System (SOF ATS). Many of these are still under development, and so are being designed with the DIS standards in mind. These should be able to optimize their DIS performance, where existing simulators may find the DIS requirements more difficult to implement. The goals of the DIS program are very aggressive and have been defined at varying levels of detail through the next several years in the *DIS Operational Concept*.

Based upon standard PDUs and network structure, DIS can be visualized as in Figure 2-14. Like the Internet, DIS is comprised of a network of networks. LANs connect a geographically grouped set of simulators, which then join a WAN along with other LANs in a similar configuration. It is probable that many of the "tightly coupled" simulators will be found among the members of the same LAN. This will reduce network traffic across the WAN. The requirements for the LAN and the WAN may differ because of the types of functions needed. Each LAN may be controlled locally and have no need for encryption, for example. The WAN may be a military network or it may be provided by a commercial entity. Here, encryption will be essential, incurring processing overhead which impacts data transmission rates.

Currently, all communication is performed via broadcast messages. The DIS committees realize that, as the number of nodes climbs from 1,000 to 100,000, this technique will not scale. Therefore, there are efforts to create methods to multi-cast packet distribution on the network. This is probably most essential in the WAN environment, and may not be necessary on the LAN's. Experiments have been performed using the SIMNET project to identify the level of PDU exchange that will be experienced in DIS. These indicate that DIS will be able to function with 1,000 entities on an FDDI or ATM network, but the multi-casting problem must be solved if the number of nodes is to grow to the 100,000 entities envisioned [Smith, 1994].

The results of one of the studies are summarized in Figure 2-15. In this study, three different categories of PDU exchange were used: ground movement, air movement, and radio communications. The last indicates an area where

Entities	Quantity	Bits/PDU	PDU/sec	Total Bits/sec	Total Packets/sec
Land					
- Manned	400	1928	3	2313600	1200
- Unmanned	550	1928	1	1060400	550
Air					
- Manned	20	1928	13	501280	260
- Unmanned	30	1928	2	115680	60
Radio Channels	60	6840	20	8208000	1200
Total	1000			12198960	3270

Figure 2-15. DIS Network Performance Analysis

multi-casting will be extremely useful. In the experiment, 60 communication channels were simultaneously operated. With broadcast communications, all of this information must be transmitted to each simulation node. The nodes then determine which of the data is useful. If the nodes were assigned to multi-cast groups according to the radio channels they were currently using, this would decrease the information received by orders of magnitude. Of the 60 radio channels in operation, a given node may be using one or two of them, eliminating 58 or 59 of the data steams propagated to that node.

Another issue being addressed is the geographic coordinates to be used in DIS. It does not matter what coordinate system nodes use locally, but each must be able to translate this into a standard form in order to communicate with others. The DIS standard is a three-dimensional geocentric system, represented as Latitude, Longitude, and Distance from the center of the earth. Many existing simulations were designed to operate in a very limited area and were therefore free to generate their own local coordinate systems. These may have been a standard Cartesian x-y-z or the Army UTM.

Unfortunately, neither is appropriate for interaction and communication with simulators operating in geographically dispersed areas. Therefore, these must be converted to the new system, or the node must use a translation algorithm in communicating

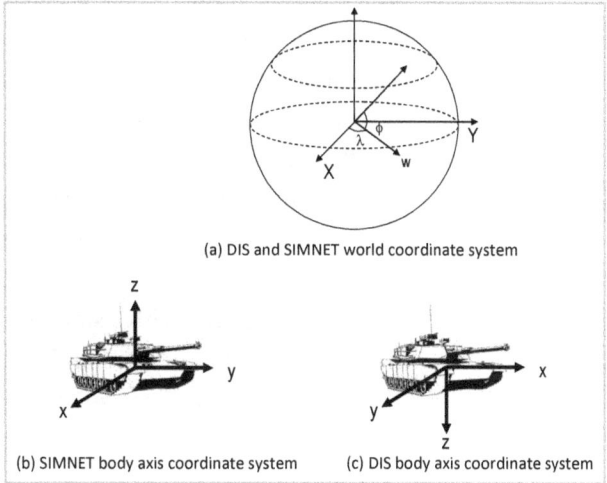

(a) DIS and SIMNET world coordinate system

(b) SIMNET body axis coordinate system   (c) DIS body axis coordinate system

Figure 2-16. Virtual Body Coordinate System

with other nodes. The latter is more common for existing simulations. The problem is illustrated quite well in Figure 2-16a-c. The DIS coordinate system and the SIMNET coordinate system are shown for entity location (16a) and for entity orientation (16b-c) [Lin, 1993].

In accordance with DIS's ambition, they intend to add interfaces with real-world entities such as troops in the field, operational aircraft, and ships at sea. The first step in accomplishing this is to assign an accurate location to each of these. This will be done by equipping each with a Global Positioning System (GPS) receiver and a transmitter. The GPS will calculate position based on orientation from the GPS satellites in orbit, an elaborate triangulation problem. This will then be transmitted to a "Central Node" which will be responsible for injecting the information into the DIS simulation network. Communications will be two-way, enabling the simulation to tell real-world participants the level of damage incurred from simulated entities and which network entities are detectable by real unit sensors. Transmitting damage/kill information will result in the same actions that occur now in the digital environment. If a soldier is involved in a field exercise using laser enhanced weapons and sensors, a "killed" signal will deactivate his rifle and signal the kill. Making simulated entities appear to real sensor systems will be more difficult. Virtual reality technologies are envisioned to cue human eyes and ears. Radar, infrared, and sonar systems must be integrated with a simulation module, which will inject the simulation information just as if it had been relayed from the sensor's antenna. This type of real-world interaction will remain in the experimental and demonstration phases for many years to come.

2.8 Computer Generated Forces Models

The DIS community has recognized that it is impossible
to acquire simulators for all of the objects found in a typical combat engagement. But, without these entities, the
environment does not sufficiently replicate events that
will occur in an actual war, reducing the value of simulated training. In fact, the ability to generate more simulated
entities than one can field physically is one of the attractive advantages of simulation-based training exercises.

Since the DIS community envisions simulations with
10,000 to 100,000 entities, most of these must be generated and controlled by computer models. These models will
be either completely independent of human interaction,
or controlled at some macro-level. These have come to
be known as Computer Generated Forces (CGF). As they
exist today, they have much in common with constructive
simulations. Where a DIS simulation replicates a single
object (tank, truck, helicopter, aircraft, etc.), a constructive model replicates thousands. CGF is an attempt to
use a single model for multiple objects, without losing
the fidelity provided in the virtual world. This attempt to
take the best of both worlds makes these model a natural
bridging mechanism for vertical integration.

The DMSO commissioned a survey of the different CGF
models and techniques being used [DMSO, 1993]. In doing this, they set four criteria that a model must meet:

1) DIS compliant. It communicates with the outside
   world via DIS PDUs.
2) Real-time interface.
3) Entity-level representation of combat elements.

4) Credible surrogate for the behavior of manned units.

With these criteria they found eight models which could be considered CGF and another five which were moving in that direction.

Comprehensive CGF	CGF Related
BBN SAF 4.3.3	IBM Blackboard Research
ModSAF	UCCATS/JCM/AMOEBA
BDS-D CGF	BBS/DIS
IFOR/WISSARD	Eagle-SAF-SIMNET
CCTT SAFOR	RESA-UVA
SWEG/SUPPRESSOR	
IST SAFOR	
Janus A	

The executive overview of the study states that several good ideas are found in the CGF architectures and should be used in future development. These are:

- Explicit representation and capture of command and control information,
- Standardized messages and formats for command and control across all echelons,
- Modular, reconfigurable entity representations,
- Distributed storage of state information,
- Arbitration schemes for resolving conflicting goals,
- 3-D view controllable from CGF operator console.

They also found notable weaknesses:

- No development structure. Do not use existing algorithms,

- No documented, robust set of requirements,
- No structure for including other projects in the development effort,
- No interface with the DIS PDU and terrain development groups.

The broadness of DMSO's criteria included models which are not all useful in our discussion of vertical simulation integration. Therefore, we will discuss only those models that offer promise in this area.

2.8.1 ModSAF

Of the CGF models BBN SAF 4.3.3 and ModSAF can be discussed as a single entity. Although each may be found in use on different projects, ModSAF is the current, growing model. The BBN SAF is an earlier baseline used in the SIMNET project (Figure 2-17) [ModSAF, 1994].

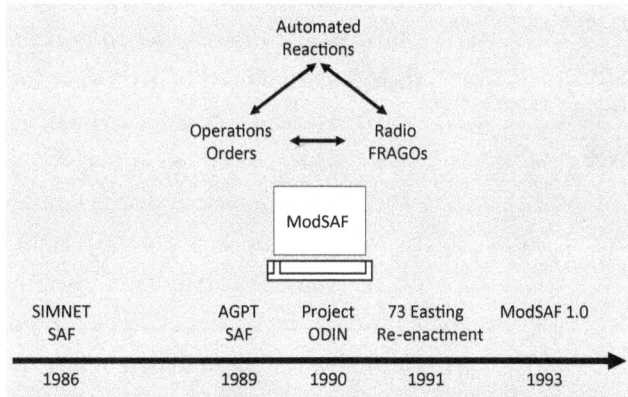

Figure 2-17. ModSAF Development Timeline

When one hears the term computer generated forces, one may naturally assume that some types of artificial intelligence techniques will be involved. This is not the case, although knowledge-based and rule-based systems are present. ModSAF does not use either of these. In fact, it is an attempt to create a user interface in which a single operator can control 50 to 150 entities. All of these entities and the

environment in which they operate, are modeled at the
DIS virtual level, so it can not be considered a constructive model, even though there are similarities. This detail
limits a ModSAF suite to tracking 1,000 total entities.

The ModSAF operator uses a 2-D graphic plan-view
display (God's eye view) showing a terrain map and object
icons. He is then responsible for manipulating these
such that they present a credible force to the virtual level
simulators they are interfaced with. Although the display
may look constructive, the models do not contain many
of the statistical characteristics of constructive models.
For example, where CBS may determine object-to-object
visibility using average terrain and Monte Carlo decision
algorithms, ModSAF deals with actual terrain points and
calculates the actual intervisibility for each encounter.
This type of difference was one of the primary causes for
the development of CGF models, rather than the adapta-
tion of existing constructive models.

ModSAF contains four distinct databases, two local and
two virtual. The local are the terrain database and the
object parameters database. These provide the beginning
picture of the battlefield. The terrain does not change
through a war, though dynamic terrain is an active research area. The Object Parameters describe the entities
that are owned and controlled by the ModSAF model.
These characteristics change as combat events unfold. The
virtual databases are those which are created and updated
by external models sharing the network with ModSAF.
These are the Persistent Object database (PO-DB) and
the DIS database (DIS-DB). The PO-DB creates the battle
scene which includes external entities. It is dynamic and
includes the contents of the Object Parameters database.

The DIS-DB contains all of the messages that have arrived from the outside world. These include those that have created the PO-DB and others involved with simulation management.

ModSAF is composed of three distinct models: SAFstation, SAFsim, and SAF-logger (Figure 2-18). SAFstation is the user interface provided to the operator. This is a 2-D map with drawing tools, terrain analysis tools, and mission planning templates. With a single SAFstation, an operator may control a single or multiple SAFsim processes, which actually model object activity. An operator is responsible for constructing missions for the entities he controls. Missions may involve multiple objects and they span a period of time, referred to as a mission matrix. Multiple matrices within the same SAFstation are referred to as a matrix stack. Once constructed, the matrices and stacks are given to the SAFsim for execution. The operator may also interactively enter commands based on the combat situations displayed on the screen.

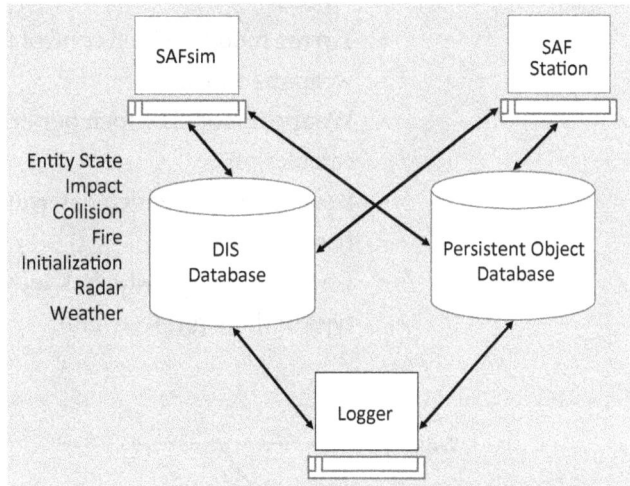

Figure 2-18. ModSAF Components

SAFsim manipulates the entities it controls based on the mission matrix stacks, interactive orders, local terrain information, and enemy unit engagements. A separate task

manager for each simulated object takes events from the
queue and executes them. Objects use avoidance algorithms
to modify their behavior to maneuver around trees and
rivers. SAFsim contains four different model libraries which
define the level of fidelity of the simulation. These are:

- **Hull** models which control the movement of the
  object,
- **Turret** models which control the aiming rotation of
  weapons and sensors.
- **Weapon** models which perform probability of hit
  calculations.
- **Sensor** models which determine whether detection
  has occurred.
- **Damage** models which determine the level and
  type of damage given a hit.

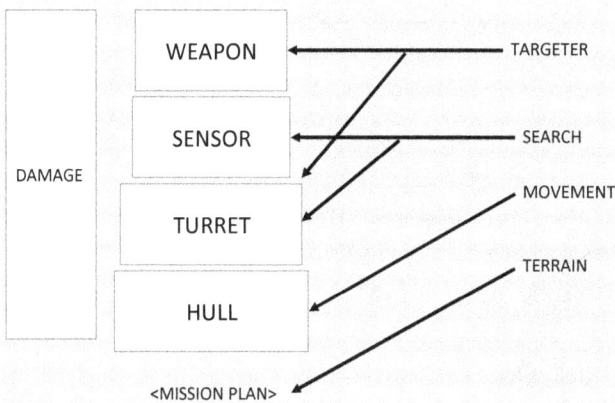

Each of these
is directed by
tasks such as
movement, ter-
rain, search, and
targeting (Figure
2-19). When ap-
propriate, objects
seek to engage
enemy units that
are encountered.

Figure 2-19. ModSAF Model Libraries

SAF-logger records all incoming information and the conditions under which decisions were made. These are used
to evaluate the models and to replay the results of combat
events.

ModSAF is a purely interactive simulation with automatic terrain avoidance and engagement algorithms. Since the locations of all entities are stored in the DIS-DB it is possible to pass control of certain objects from one SAFsim to another under the same SAFstation. This creates a method of load balancing and fault tolerance which makes the system attractive. ModSAF is designed to operate on a 2Hz update cycle when fully loaded. This does not comply with the DIS 100–300 millisecond standard for manned entities. The impact of this is not currently known.

Consider the following:

Model	Objects Controlled
DIS Simulator	1
ModSAF	50 - 150
ALBAM	400
TACSIM	10,000
CBS	20,000

A quick glance makes it clear why some type of bridging model is needed between DIS and ALSP, and why ModSAF is considered for this function. In a vertically integrated exercise, given that DIS may field 1000 simulators (very optimistic), 19,000 must be controlled by ModSAF. Assuming that those 1,000 DIS objects are geographically grouped, and therefore encounter at most half of the CBS objects, this is 9,500. As currently designed, this will require 64 ModSAF simulation suites, which is a feasible number for initial vertical interface experiments.

2.8.2 IST SAFOR

The IST SAFOR was commissioned as a low cost system. It resides on an 80386 based PC, using the MS-DOS operating system. Promising work has been done to perform terrain reasoning and weapon selection for the controlled entities. But, it is designed to control only a single object. As a system, it can not be used for vertical integration, although the algorithms and model architecture may be employed in future CGF systems.

2.8.3 IBM Blackboard

The IBM Blackboard Research is based on artificial intelligence (AI) techniques. A single, common blackboard is used to record and modify information from several different sources. Units and events are monitored by AI algorithms, which use these to make decisions. Decisions are then translated into actions for locally controlled objects, and commands are issued to carry these out. Experiments have been done with two platoons of M3 Bradley vehicles and Soviet BMP's, but the results are not published in the DMSO report. Although just a research project, it aims for the capabilities needed for a vertical interface.

2.8.4 Rasputin

Although not included in the DMSO survey, the Rapid Scenarios Preparation Unit for Intelligence (Rasputin) has the ability to perform CGF functions. This system was built to provide a simple graphic interface in constructing a scenario, including timed movement patterns. Commands are issued at levels of organizational aggregation, such as battalion headquarters. The stated objective is simply to move from point A to point B. Lower level objects are then constrained to follow realistic courses through the terrain database by a rule-base of military operations. This requires that they follow roads, avoid obstacles, and optimize their paths. A knowledge-base containing thousands of rules is used. Unfortunately, the system can not provide real-time calculations of these steps. It is meant for engineering-level studies. The exact time required to calculate movement is not generally known to the simulation community.

The DMSO study identified four enabling technologies necessary to create true Computer Generated Forces.

1) Behavioral representation of decision making.
2) Processing and interpreting terrain information.
3) Exercise support (scenario generation, plan development, and live analysis).
4) General computing technologies, to include processing speed, parallel processing, networking, and computer image generators.

As these technologies develop, the ability to truly generate and control realistic units via computer will grow.

2.9 Vertical Integration Prototypes

There are currently several experimental projects being conducted on the viability of connecting virtual and constructive level simulations. All attempt to join one virtual and one constructive simulation, with a CGF model as the bridge. The first step is to disaggregate a constructive unit, such as an armor company, and instantiate the objects of that unit in a virtual simulation. The second step is to assign virtual attributes to those disaggregated objects. The third step, which is currently missing, is to automate the operation of those objects in the virtual world.

2.9.1 AWSIM to ModSAF

The Air Warfare Simulation (AWSIM) is one of the members of the ALSP confederation. It is responsible for controlling fixed wing aircraft and surface-to-air missile batteries [Summary Report: The Ninth ..., 1993].

The Modular Semi-Automated Forces Model (ModSAF) is the leading CGF model in the DIS community. The first capabilities developed in the model were air-to-air and air-to-ground operations.

These two models are being joined by the MITRE corporation as part of the ARPA Simulated Theater of War (STOW) 94 demonstration. The requirement is to use AWSIM to generate and control aircraft sorties in support of a virtual level exercise. To accomplish this, the combat area has been divided into two sections. The AWSIM/ALSP box is the large area and completely contains the smaller DIS/ModSAF area (Figure 2-20).

AWSIM controllers will generate missions in response to support requests from the exercise participants. These missions will be launched from AWSIM air bases, complete with the appropriate number of aircraft, fuel, munitions, sensors, a flight path, and an objective. Once launched, the AWSIM flight model controls the mission until it enters the ModSAF play box. At that point, control is handed to ModSAF and the mission becomes a ghost unit in AWSIM. As a ghost, the mission will not be attacked by AWSIM air defenses or enemy aircraft. This responsibility now rests upon the networked DIS simulators. AWSIM now tracks only the general location of the mission, in preparation

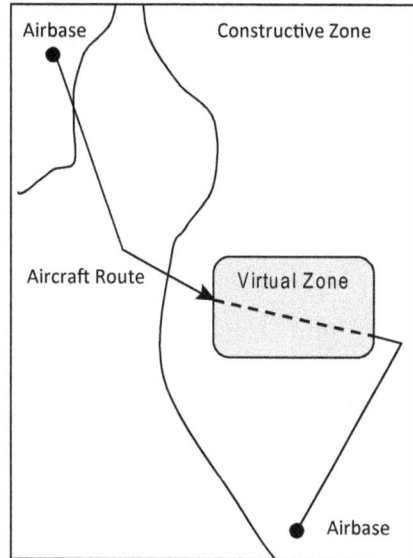

Figure 2-20. AWSIM to ModSAF Functionality

for receiving control back from ModSAF for after action analysis. Once the aircraft exit the ModSAF box, control returns to AWSIM in order to complete the mission. While in AWSIM, aircraft must operate in the electronic warfare, air defense and air-to-air combat environment native to the model. This allows AWSIM to widen the scope of the scenario being replicated in the DIS-level STOW exercise.

While in the ModSAF box each aircraft is individually instantiated and controlled by the ModSAF operator. This task is man-power intensive and will require several ModSAF operators for each AWSIM operator generating the missions. Since AWSIM is operating in a support function, independent decision making may be exercised in mission composition, arming, and flight path generation, based upon objectives supplied by the DIS commanders.

To enable this technique, ModSAF was joined to ALSP as a quasi-member of the confederation. Special messages and algorithms had to be created to enable this, since ModSAF does not posses all of the characteristics of a constructive model, nor are ALSP and its members the driving force of the demonstration exercise. This structure will test the usefulness of constructive simulations in enhancing virtual exercises. But, the ultimate goal of vertical integration is to allow both levels to participate equally in a simulated environment.

2.9.2 JPL Alpha Project

The NASA Jet Propulsion Laboratory has created a research effort dubbed the Alpha Project. This is an attempt to bridge the constructive-virtual gap, using mathematic algorithms, rather than man-in-the-loop simulators such as ModSAF. Although in its infancy, the project has begun to define splines for continuous flight-paths from the discrete mission points given in a constructive simulation. They are also defining sampling criteria for replicating a mission flight profile from a continuous DIS flight-path

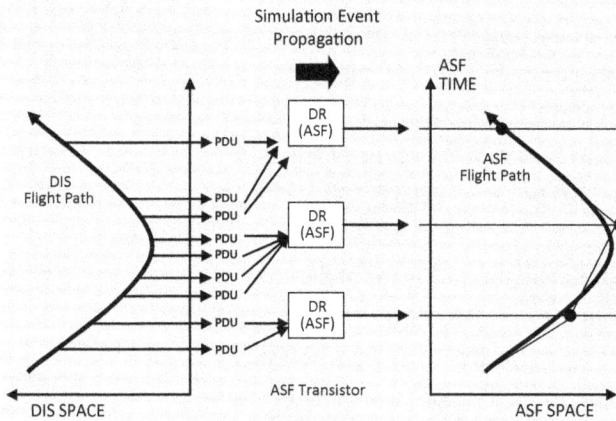

Figure 2-21. JPL Project Alpha Links

(Figure 2-21). A DIS flight-path is actually a discrete set of points, just as in the constructive world, but these points are defined to be 100-300 milliseconds apart. Should an object not deviate from its last orientation, dead reckoning (DR) algorithms are used to project its position along the last known vector. Therefore, the sample is made to meet these criteria and minimize network data traffic.

The project has defined the constructive piece as the Advanced Simulation Framework (ASF) in the Figures. This framework may represent enhancements to existing simulations. such as AWSIM and CBS, or an architecture to be used in constructing the next generation of constructive models. Though the later is the most desirable, budgetary constraints and project demands will influence the development and implementation of this research project.

2.9.3 Eagle to BDS-D/SIMNET

Eagle is an Army TRADOC Analysis Command (TRAC) model for constructive corps and division level combat operating on a Symbolics and a Sun workstation. Units are represented down to the company and battalion level. It is used to conduct combat analysis studies for evaluating weapons systems, command and control, military doctrine, and the effectiveness of organized forces [Hardy, January 1992].

Battlefield Distributed Simulation—Developmental (BDS-D) is a program designed to bring multiple virtual simulations together in a shared environment. The Simulator Networking (SIMNET) project is a group of armored vehicle simulators that have been networked into a shared virtual environment. The actual interface with the constructive world is via the Institute for Simulation and Training's Semi-Automated Forces model (IST SAFOR). This is a PC based model comprised of an Operator Interface (OI) computer and a simulator computer. As described in the CGF section, IST SAFOR primarily performs terrain reasoning and some weapon selection decisions preparatory to combat. All other operations are handled by human operators.

There are several similarities between this project and the AWSIM to ModSAF project, although this is more ambitious. It attempts to maintain a simulated environment in which both the virtual SAFOR and the constructive Eagle participate equally (Figure 2-22). Eagle is not just a support tool for a SIMNET exercise. In joining Eagle and IST SAFOR, every effort was made to maintain the independent operating capability that each started with.

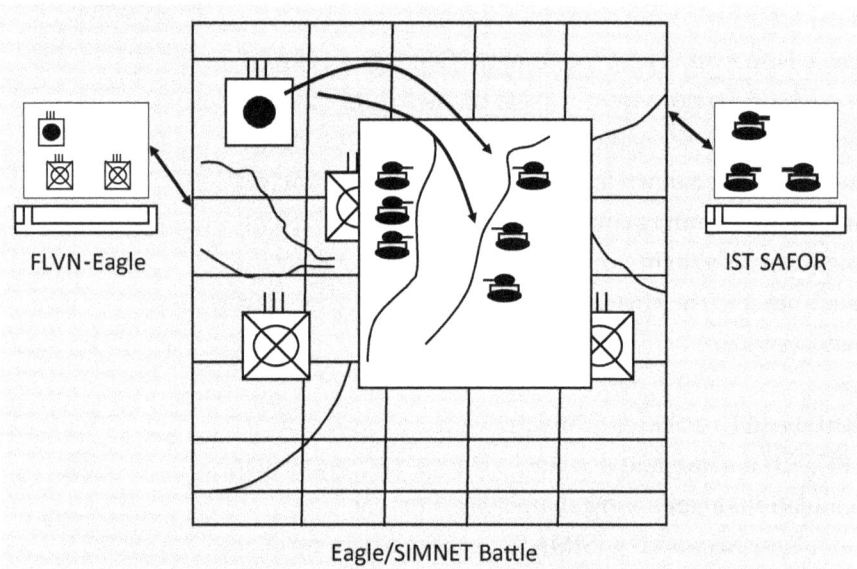

FLVN-Eagle

IST SAFOR

Eagle/SIMNET Battle

Figure 2-22. Eagle to BDS-D Functionality

The Eagle play box is a 100km by 100km square. Within this box, are one or more smaller CGF areas where IST SAFOR operates. Each CGF area is controlled by one or more IST SAFOR models. The number required is completely dependent upon the number of vehicles operating within the area. Unlike AWSIM/ModSAF, an Eagle unit does not disaggregate upon entering the CGF area. Disaggregation points have been defined within the CGF area. These form gateways for transferring control from Eagle to IST SAFOR and back (Figure 2-23). The Eagle and SIMNET worlds are parallel layers like a multi-dimensional chess board. The disaggregation points are the only portals for traveling between these layers. Multiple CGF areas can overlap and the IST SAFOR model receiving the Eagle unit is determined by the portal that is entered.

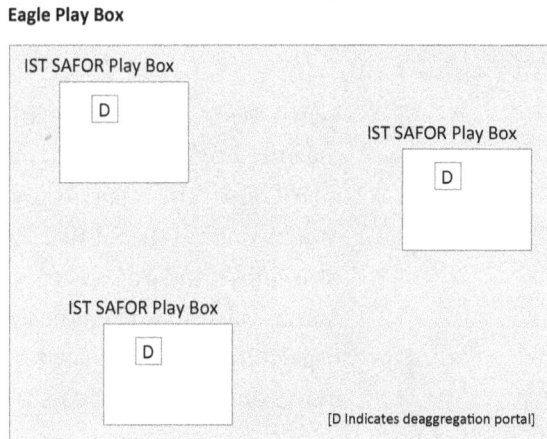

Figure 2-23. Eagle to IST SAFOR Gateways

When an Eagle aggregated unit enters one of the portals to CGF, unit control is passed to the IST SAFOR model. Within Eagle, the unit no longer moves, although it can still receive operator orders. Once transferred to IST SAFOR, an Eagle unit is disaggregated and the objects begin to appear in the SIMNET world connected to the CGF models. This magical apparition is a characteristic that must be eliminated at some future point in the project. The vehicles are now controlled by the IST SAFOR operator, with the aid of the terrain reasoning algorithms.

The manual work-load, together with computer processing limits determines the number of objects that can be controlled in each CGF model.

Unlike AWSIM/ModSAF, when an Eagle unit has turned control over to the CGF, it is still vulnerable to other aggregated units operating in Eagle who have not entered the portal. The unit's position continues to change as the CGF objects move. The location is the center of mass of the CGF objects. Enemy objects detected by Eagle objects within the CGF are also registered as detected in the constructive model and appear on the constructive operator's screen.

When the Eagle operator wishes to regain control of the unit, he simply issues a movement order which will take it out of the disaggregation portal. The objects then disappear from the SIMNET world and the aggregate unit continues standard constructive operations within Eagle. A unit will also be reaggregated out of the CGF back into Eagle if the center of mass of the disaggregated objects moves outside of the transfer portal.

In addition to the above capabilities the developers have created the ability to receive indirect fire within the SIMNET world from an Eagle artillery unit (whether constructive or disaggregated into the CGF). The CGF operator sends a request for fire to the Eagle operator, who manually enters the order directing the salvos into the CGF area. The shells enter in the SIMNET world creating visual explosions and actual damage. The damage may effect SIMNET objects and disaggregated Eagle objects.

The developers of the Eagle/BDS-D interface report that it is possible to operate at two levels of fidelity. In the

first, disaggregated units in the SIMNET world appear as
individual vehicles. In the second, Eagle units appear as a
single constructive icon in the SIMNET world. This provides a level of integration which does not require large
amounts of network traffic.

In the above discussion, one point that needs to be clari-
fied is that Eagle units disaggregate into the CGF, where
they may engage other CGF units, whether organic to the
CGF or disaggregated from Eagle. The disaggregated units
may appear visually in a networked SIMNET environment.
But they do not exist there in any other way. They may not
engage with other SIMNET objects to damage them or be
damaged by them. The appearance capability is provided
only in order to create a realistic "looking" scenario in
SIMNET and to stimulate SIMNET sensors.

2.9.4 BBS to SIMNET

The Brigade/Battalion Battle Simulation (BBS) is used
to train brigade and battalion commanders in combat
decision making. The commanders operate in a field
command post from which they issue orders based on
the battlefield they see on their maps. These orders are
received by soldiers operating BBS computer terminals
rather than by their real-world equivalents in the field.
The combat events are then executed in the BBS model
and the results relayed back to the command post. BBS
is very similar to CBS, but with a slightly higher level of
resolution. BBS operates on a 15 second time-step, CBS
on a 1 minute time-step. BBS is a networked simulation.
Multiple company-level nodes are connected together to
simulate a brigade or battalion. Each object is instantiated
at a unique location. These characteristics make it similar

to the network of object simulators in the virtual world [Summary Report: The Ninth ..., 1993].

SIMNET and the SIMNET SAF have been defined earlier in the book.

This integration project has proceeded much as the Eagle to BDS-D interface did. The primary method of joining the constructive and virtual worlds is to insert a CGF model between the two and build interfaces to it on both sides. The CGF model used for this project is known as the SIMNET SAF. This is the original Semi-Automated Forces model developed to support SIMNET exercises between networked tanks and helicopters. Therefore, the interface between SIMNET and SIMNET/SAF already exists. Since both the constructive and virtual pieces of the system existed, all work has been focused in the intermediate level.

During phase one of the project the goal was to create the appearance of BBS controlled units, disaggregated by the SAF, in the SIMNET world. These unit/objects were then to move according to orders given by the SAF operator and the characteristics assigned by the BBS operator. This has been accomplished.

The goals of phase two were to create a functional prototype in which units and objects could interact in active combat. This too has been accomplished. What remains is to expand the degree or interaction and the sharing of events. The "BBS/SIMNET Functional Validation Test Report" describes the results of the phase two test. Although conceptually a test, the integration is limited by several design characteristics of the BBS, SAF, and SIMNET models.

The decision to disaggregate a BBS unit into the SAF and SIMNET world is based on its proximity to a SIMNET object, rather than by entry into geographic transfer areas. When a BBS unit is within 5km of a SIMNET object, a message is sent to the AIU informing it that a disaggregation must be performed. This results in the AIU preparing the data required by the SAF to instantiate all of the objects of the BBS unit in the SAF/SIMNET world. The SAF engine receives this data and creates the appropriate objects. These are then visible and active both in the SAF and in SIMNET.

The AIU manages the differences in time progression in the virtual and constructive world. BBS time progresses in 15 second intervals. SIMNET time is continuous. Therefore, unit movement within each BBS time-step is buffered in the AIU and the final unit state is released to BBS at the beginning of its next time-step. If SIMNET movement does not deviate from the last reported heading and speed, there will be no new messages for the units position. Under these conditions, the AIU calculates the dead reckoned (DR) position of the objects at the end of the current BBS time-step. Since the ultimate goal is to join constructive simulations with a DIS standard network, rather than a SIMNET standard, the AIU was designed to be able to translate BBS messages into either DIS or SIMNET packets. This will allow migration to the DIS network at some future point.

The SAF engine creates and deletes objects according to their relative locations between the BBS and SIMNET entities. SAF is responsible for moving units according to the last supplied orientation and speed. It must also position the objects of a disaggregated unit according to

the BBS supplied "firing posture". This posture indicates a
template for the formation of the objects belonging to the
unit.

The Simulation Controller (SIMCON) is responsible for
providing some of the automated functions available in
BBS, but not in SAF or SIMNET. These include the effects
of air support, combat engineering, resupply, and automatic responses to combat losses (defensive positioning
and withdraw). These functions provide a realistic looking
scenario in the virtual world, but are functionally inert.
For example, minefields can appear in the virtual world,
but they will have no effect on objects driving over them.

In this project, the SAF has been designed to model
operational units, rather than simple unrelated objects.
Since the relationship between objects is maintained, this
allows response by the entire unit to external events. One
instance of this particularly impressed observers at the
validation test. A flight of two helicopters passed in front
of an air defense battery. One of the helicopters was shot
down. In response to this, the other helicopter dropped in
altitude and turned to engage the ADA battery. Had each
object been maintained without the relationship of "2
ship flight", the surviving helicopter would have continued
on its original course and would have been destroyed as
well.

BBS/SIMNET has also implemented the ability for constructive artillery to fire into the virtual world and cause
damage.

It appears that this project has made an excellent choice of
two simulators to be interfaced. The networked nature of

BBS, and its 15 second time-step, make it more compatible with the DIS world than either Eagle or AWSIM.

The approach used by the AWSIM/ModSAF project is particularly forward thinking. They realize that the constructive world will be joined by the ALSP and have tried to use this structure in their vertical integration. The BBS/SIMNET interface has begun with, what seems to be, a natural marriage or constructive and virtual models. This may avoid discontinuity long enough to provide a bridge, without some of the compromises that would otherwise arise in the design.

We would like to point out that all of these vertical prototypes are man-power intensive. During an actual exercise, this man-power must be taken from the units being trained. This is one of the problems with constructive simulations today, they require that only part of a unit receives realistic training, while others members are supporting the computerized infrastructure. Adding the load of maintaining a vertical bridge may exceed the ability of military units to operate with the staff remaining to them.

Translational Functions

In order to construct a vertical gateway between simulations, we must first identify the type of data that must be passed across it, the environment in which it will be used, and the translations that this implies. In considering these we will use an example in which a known aggregated unit in a constructive simulation is communicating with the known objects that make up such a unit in the virtual world. We will consider first the cases of translating from constructive to virtual, and then from the virtual to the constructive.

One of the first issues to arise is whether considering a single constructive simulation and a single type of virtual simulator is sufficient. The answer to this will unfold as we explore the issue of unit control versus representation. Since the DIS and ALSP projects are attempting to force a standard on simulation integration, the problem should not be impacted by the type of simulation we choose to integrate with.

In creating the vertical bridge, we are making assumptions based on military doctrine, experience, and our understanding of the uses to which the simulations are being put. This results in a loss of human generated unpredictability.

3.1 Relationships

The first step is to determine the relationship between the constructive entity and the virtual objects. It is common to aggregate into military units such as companies, battalions, and brigades. The military has the very useful characteristic of defining the composition, operational characteristics, command structure, and options available to these units. In theory everything a soldier does in the field can be traced to a manual which was written to train millions just like him to behave in a predictable manner. This characteristic is absolutely necessary in order to create a fighting force that can be directed with the least amount of confusion and variation.

3.1.1 Templates

Doctrinal behavior allows us to build templates which are software instantiations of the rules described in the military manuals. These should then mirror the desired behavior of the military unit.

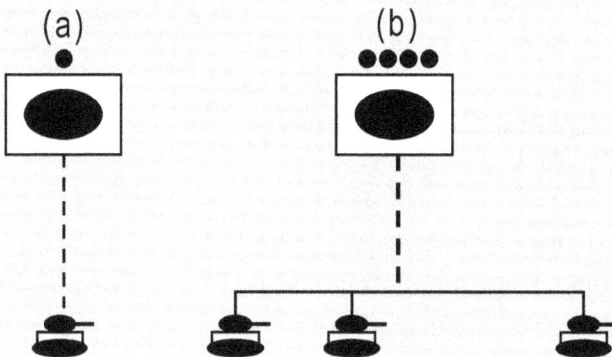

Figure 3-1. Constructive-Virtual Unit Relationships

A trivial case is that in which a single object, say a tank, is aggregated into a constructive model. This is shown in figure 3-1a. In both levels of simulation the icon displayed represents the same piece of equipment. But, once it begins to operate, it with be acted upon by different models of the environment. A more complex, and more common, case is shown in figure 3-1b. Here, a tank company is composed of various pieces and numbers of equipment. These may include tanks, trucks, jeeps, humans, and small arms. The problem of maintaining consistency is much increased.

To define the placement of the objects in the complex unit we must determine its type, size, and activity. These three characteristics will provide a basic template for operational deployment. Our unit is an ARMORED, COMPANY, in CONVOY. This implies the shape of the unit as being in a road march, the equipment arrayed in columns proceeding down a road. It also places the command tank at a certain location in these columns.This basic deployment will then be acted upon by characteristics of the environment - the width of the road, the proximity to the enemy, the time of day, the mission objective, and the surrounding terrain. Again, we can rely on doctrine to give us the perfect pattern in which to array our pieces of equipment. If several tank companies were to be disaggregated from a larger battalion, the method of deducing the deployment would be essentially the same.

In discussing the deployment pattern, we are actually defining the physical location and orientation of the objects. These are some of the primary characteristics needed in describing an object. Assuming that the aggregate unit also knows the types of objects it contains, a mapping can be made which identifies the virtual objects that match these. This provides the appearance characteristics often needed in the virtual world.

3.1.2 Dynamics

We have hinted at the fact that the unit will not exist in
a vacuum, but in an environment which will have some
impact on the template so neatly laid out in the military
manual. Several of these modifiers have been listed, but
their exact effects have not been defined. It is not our
intent to enumerate them here. We merely wish to point
out that the modifications have also been distilled from
centuries of lethal combat situations and placed in manuals in an attempt to make current forces more survivable
and effective than their predecessors. In addition to the
printed word, we may modify the deployment patterns by
information gathered from military experts and Monte
Carlo methods. The latter are used to introduce that
amount of random variation found in all human activities. This imparts an unpredictability that makes the
units appear to be even more human. Rules are needed to
bound the randomness and prevent the deployment from
becoming chaotic.

Given that the aggregate unit is disaggregated into the
virtual world, does the fact that environmental factors entered into the locations of the vehicles have any impact on
the location or orientation of the constructive unit?This
depends upon which simulation is actually in control
of the objects/units we are considering, rather than just
representing them. This will be explored further in a later
section.

Viewing the above situation from the virtual side, we see
a different problem. Given the locations of a set of objects
known to belong to a specific constructive unit, what is
the derived location of that unit? The first to answer this

question will respond that we should average the locations
of the virtual objects to arrive at the constructive location,
and perhaps some type of weighted average can be used.

Unfortunately, this answer is too simple to suffice.
Remember that the constructive to virtual transformation
was accomplished via a template based on a defined command point. Averaging virtual locations will certainly not
arrive at the same location of the aggregate unit that may
have been used to lay-down the virtual objects, originally.
A more accurate method is to appoint one of the virtual
objects as the command element. Its location will then be
used as the location of the constructive unit. This scheme
will produce a more realistic and consistent representation of the forces. The constructive location is typically
determined by the location of the command element as
reported to upper levels of command. This object/loca-
tion may not even coincide with the actual location of a
physical piece of equipment. The command object may be
the instantiation of a unit characteristic in object-form to
produce the desired effect.

Even using this method, the location of the constructive
unit that is derived from the virtual objects may not be the
same as that of the original constructive unit from which
the virtual objects were disaggregated. Remember that the
virtual locations are all effected by the virtual environment. Since the details of this are available only in the
virtual realm, it is impossible for the constructive model
to take these into account in placing the constructive unit.
Given that the mission is to join disparate simulations,
this is an artifact that can not be avoided.

3.1.3 Consistent History

Given that a complex unit has disaggregated into the virtual world, the virtual objects will begin to experience unique individual changes. Virtual models will move these objects into new formations not totally consistent with the locations that would be derived from templates, doctrine, and environment factors. Certain vehicles will be destroyed or damaged. Different equipment types will consume supplies, such as fuel, at different rates. These types of changes create the dynamic, variable battlefield that is valuable for the trainee experiencing the virtual world.

At some point, the virtual objects may be aggregated into the constructive unit. If these then lose their individual identities, there will be no data structure to maintain the unique histories of the objects. If this data is lost, the virtual exercise becomes a sequence of unrelated events, depriving the training audience of the consistency present in actual combat. For this reason, it is necessary for disaggregated unit information to be maintained, even when the "official" unit is re-aggregated. Uniqueness will then be maintained and reactivated at the next disaggregation of the unit.

The virtual history being stored need not be updated during the constructive control period. This will occur when the unit is again disaggregated. This technique is similar to computer memory caching, where the actual data value in long-term memory is not updated while the information is stored in the cache. The cache value is changed and the operating system uses this value as long as it remains in the cache. The long-term storage value is only changed when the information is swapped out-of the cache.

The implications of this scheme are: 1) increasing variability in the virtual representations as time progresses, and 2) reduced disaggregation calculations as time progresses. The first occurs as more of the identical constructive units experience and store virtual variations. These units become more and more heterogeneous. This fact should increase the realism found in the virtual world by reducing the predictability inherent in dealing with "cookie cutter" units. The second is a by-product of storing past disaggregation calculations. This assumes that it is less expensive to update a stored history state than it is to calculate one for the first time. Although location relative to the command object need not be calculated, each object must be founded and its individual information passed to the appropriate virtual object.

Assume that a disaggregated, constructive unit has not changed position, orientation, posture, or any other virtually represented characteristic. Reaggregation is caused by events in the virtual world, such as all other units departing the area. The next disaggregation, perhaps caused by another virtual object reentering the area, will require little or no update to the virtual objects.Although changes may occur from events like the consumption of fuel, major computation expenses are avoided.

An update can consist of, at most, the same calculations needed to perform the original disaggregation. The difference is in the search required to locate the appropriate object to update. If the search is faster than the instantiation of memory and data structures, updates will always be faster than disaggregation.

3.1.4 Control

The final characteristic of the relationships between the
constructive and virtual units is the same as that between
any networked simulations, control. Although many simulators may represent a unit in their domain, only a single
simulation can be allowed to control a unit's behavior.
This is not to imply that every one of the unit's behaviors
must be represented in the same simulator, but rather
that each behavior is controlled from only one place. The
movement of the unit, for example, may be controlled by
simulation A, but the performance of its sensors may be
controlled by simulation B. Others may represent other
characteristics, but all requests to change them must be
sent to the appropriate simulation to be acted upon or
rejected.

This characteristic is necessary to avoid creating impossible events in different simulations. For example, if a
constructive and virtual simulation were each allowed
to update the location of a unit in response to a single
movement command, the results would not be the same.
The constructive simulation would perform movement
which may be impossible, given the constraints detailed
in the virtual world. These constraints would certainly
result in an arrival time difference between the virtual and
constructive units at a given destination.

3.2 Actions

We will explore several of the major types of action that
military units engage in. These will illustrate the rules and
techniques that must be used in maintaining a joint vir-
tual/constructive simulation exercise.

3.2.1 Movement

The first action is the most basic and most common,
movement. Let us begin with our simple unit, the
aggregated single tank. This unit/object, represented in
both the constructive and virtual worlds, is ordered to
move from point A to point B (Figure 3-2). Since the tank
in the construc-
tive simulation
experiences
constructive
terrain and
obstacles, the
time for it to
move the
distance must be
greater than if it
had moved across

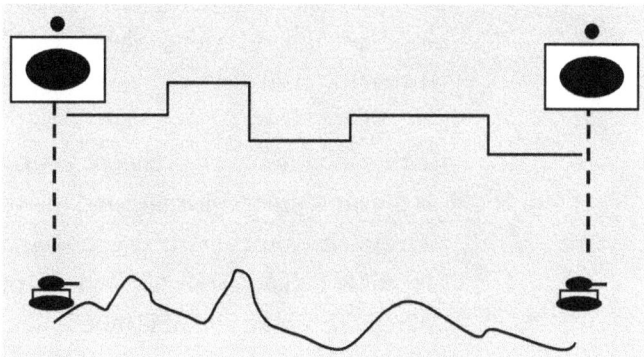

Figure 3-2. Simple Unit Movement

a smooth, table-top like surface. The terrain in most
constructive simulations is a form of virtual data that has
been pre-processed into convenient terrain factors
assigned to a tiled surface. The unit then experiences
degradations of its speed based on these factors.

The virtual tank is also moving across a discrete representation of terrain. The tiles in the virtual world are smaller,
but unless a continuously changing surface is used, the

effect is the same as that for the constructive terrain. Since the two terrains are both discrete and different, the time required for the tank to traverse constructive A-B will, in general, be different from that to traverse virtual A-B. Assume that the constructive pre-processor is very accurate in producing life-like tiles. This may create a situation in which the time and distance experienced constructively is equal to that experienced virtually. But, we may vary the destination point enough to create a path across one different virtual tile, while remaining invariant constructively. Only when the new virtual tile traversed is identical to the old tile will the virtual traversal time remain the same as the constructive traversal time. For all other cases, the virtual time will become greater than or less than the constructive time. It is a special case when any constructive traversal time is equal to the equivalent virtual traversal time.

In this situation, we can choose to treat one of the models as the movement controller and the other as the shadow. If we choose the constructive model, the virtual unit will appear to behave erratically and perhaps violate physics. When the terrain contains mountains, lakes, rivers, and other terrain features, actions by the constructive unit shadowed in the virtual world will cause the virtual object to drive across these features as if by magic. On the other hand, if the virtual object is the controller, the construc-tive unit will appear to behave smarter than expected, given the terrain data it is experiencing. It will avoid lakes while its constructive-only brethren move blindly across them. This unit will behave more realistically, not less. The unit will also arrive at its destination at a time more accurately representing that experienced in the real-world.

It is clear that in this situation the existence of the virtual connection dictates that it be the controller in order to maintain realism in both worlds. Since the constructive unit is not experiencing the detailed terrain in the virtual world, it will not be travelling the circuitous route of the virtual unit. It will simply traverse its route a bit more slowly. Each time the virtual unit transmits its location to the constructive unit, the constructive unit will experience a straight-line path between the two update points, while the virtual unit will have moved a further distance by negotiating its more complex terrain. Therefore, the shortest distance between two points is the constructive distance. And the fastest unit on the battlefield is the virtual unit.

If it is true that all movement must be calculated in the virtual world, then other effects are implied. Since constructive movement is being shadowed from the virtual world, constructive events occurring to the unit during its movement must impact the unit's movement. Should the constructive unit be engaged by another constructive unit, which does not have a virtual counterpart, how will this event be reflected in the unit's movement? For this to impact movement, the opposing unit must be reflected in the virtual world. There, it can engage the virtual unit and impact its movement. If this is not done, the constructive opposition cannot be engaged, since it is not allowed the freedom to position for the event; and, should it be hit, it is not allowed to effect the virtual unit's movement.

This implies that the virtual world must become the domain of all events when even one of the participants is moving in the virtual world. This effect may be localized and bounded, but, where it is a factor, the virtual world must dominate. Methods for bounding the effects of a virtual unit on the battlefield will be explored later. The constructive world may be an aggregation of the virtual, but the virtual is more than a disaggregation of the constructive. In fact it has become the master.

Consider the movement problem with a more complex unit, say an armored company (Figure 3-3). Here, the differences in event outcomes are magnified. In fact, the virtual endpoint may be significantly different from the constructive endpoint. How can differences of this type be reconciled? Either by magical intervention or by giving control of the operation to the virtual world, making the constructive world a shadow.

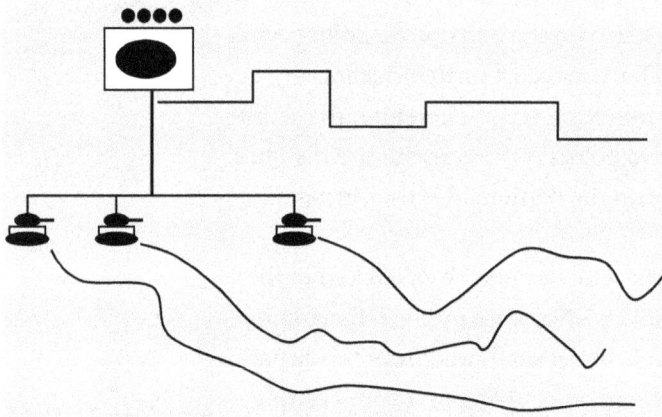

Figure 3-3. Complex Unit Movement

3.2.2 Detection

Combat simulations depend heavily upon their ability to detect the units or objects that they interact with. Determining whether a unit is detected is usually a function of the following items:

- Sensor,
- Sensor Platform,
- Target,

- Environment, Terrain, and
- Clutter Generators.

In the constructive world, these items are constructive versions of their real-world counterparts. As a result the determinations made are less than exact, but a well designed model will produce accurate results across large sample sizes. In the virtual world, these items are modeled more closely to the real system. The accuracy is still best over large sample sizes, but the deviation is smaller.

For our simple constructive tank to visually detect a target, that target must be within range of the sensor, have an unobstructed line-of-sight, be in a clear environment, and be large enough to register on the sensor (Figure 3-4). Since the terrain and atmosphere are constructive, the line-of-sight may not be totally accurate. The altitude of each

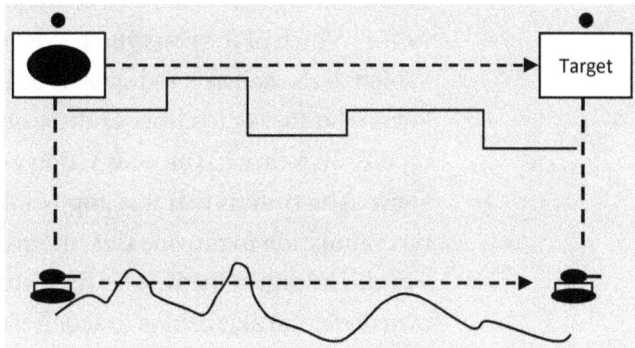

Figure 3-4. Simple Unit Detection

tile along the vector between the sensor and target may have been raised or lowered by the construction process. But, over many detections, the increases and the decreases will counter balance one another, making the exercise outcome generally fair and accurate.

If a constructive tank detects a single-object constructive target, the information can be passed to the virtual model with little misunderstanding. However, the virtual terrain may preclude virtual detection. We have essentially given the virtual sensor x-ray vision to see through the obstruction. This is hardly a realistic, instructive situation to train soldiers in. On the other hand, if the detection is handled by the virtual sensor, the results will be acceptable in the constructive world. When virtual detection does or does not occur, the constructive world cannot tell the difference between that and a Monte Carlo effect involving the constructive terrain. Again, we see that when the virtual representation is available, its results facilitate consistent integration of the constructive and virtual worlds.

Consider the detection problem when the constructive units are more complex. The detection of a complex constructive target is difficult to translate into the virtual world. Which of the constructed vehicles was detected? Which of the constructed sensors acquired the detection? When the answer to either of these questions is simplified to "all", we return to the x-ray vision problem described above. The truth is that it is impossible for the constructive simulation to provide this information to the virtual world. The only remedy is for the virtual world to run its own detection algorithms to acquire its targets. When this is done, there is no advantage to using the constructive algorithms to perform the same function. It is more efficient and consistent for the virtual world to construct its targets and pass them to the constructive world as one of its native detections.

Another problem arises because virtual time and constructive time are proceeding at different rates, at what time does the constructive detection become a virtual detection? Should this be done immediately, interrupting further constructive processing? Or should the detection be withheld until a more convenient time? One constructive time step may be one minute or fifteen minutes, this translates into 30 or 450 virtual time steps. In the virtual world, where reaction time is essential, delaying detection can be deadly.

When the units are controlled by the virtual simulation, this problem becomes moot. The virtual simulation can pass the data immediately, or hold it until the end of the current processing set. Only in rare cases will this result in a different outcome in the constructive world. An overwhelming majority of times, the event represented in the constructive world will be identical, regardless of the virtual information-handling scheme. It is clear that, in the case of detection, when a virtual entity is involved, these events should be handled by the virtual simulation and shadowed in the constructive simulation.

3.2.3 Communications

The virtual/constructive problem in communication events is very similar to detection events. Both are an exchange of information via energy. The only difference is that the energy is generated by the object (communications), rather than reflected by it (detection).

In the military realm the simple unit and complex unit examples are identical. Individual objects are not allowed to communicate with other units. These actions are reserved for the command element, who communicates with other command elements. Therefore, there is seldom a complex case. When any object is allowed to transmit and receive communications, the command structure has broken down, and the unit may be considered to be on the verge of breaking itself apart as well.

3.2.4 Posture

The deployment posture of a constructive unit is a summarized statement of where the objects would be located if they were to be represented individually. This characteristic is used to enhance the sensor and combat models to provide a richer variety of outcomes.

It does not make sense for a simple unit to have a posture. Its relationship to itself can not change. This is true constructively and virtually. When a complex unit receives a "change posture" order, it makes the change immediately. No time is involved in moving from the old posture into the new. In order to operate with a virtual simulation, this order will have to be translated into virtual orders to move to specific locations. This movement must then be performed over an appropriate time span. The appropriate virtual time to execute the order is a function of the virtual environment. The constructive simulation cannot determine this time without itself becoming virtual. Therefore, the movement must be executed in the virtual world and shadowed in the constructive. The virtual execution time is accurate for the constructive world.

Although the transformation of the order is unique, the action taken by the unit itself is identical to the movement problem considered earlier. The posture represents a known template of positions. Movement orders are derived directly from the templates and entered as movement requests in the virtual world.

3.2.5 Engagement—Direct

Engagements are typically divided into three types: direct, indirect, and precision-guided. Direct refers to weapons that require some detection of the enemy to be useful. These include tanks, small hand-held missiles, and rifles. Indirect refers to weapons like artillery, long range rockets, and unguided air delivered weapons. Guided weapons, such as cruise missiles, guided missiles, and man-steerable weapons, are typically considered as separate entities from the delivery vehicle. The methods used for these are a combination of the direct and indirect methods.

Engagement is often the result of detecting the enemy. Using our simple tank unit and assuming that the detection was accomplished using the methods described above, we must decide whether to perform the engagement in the constructive or virtual world. Although an engagement has characteristics of the detection problem, many of the confusing factors have already been incorporated into the situation. When the sensor is similar to the weapon, requiring primarily line-of-sight, the only new factor is the range of the weapon. Each model can decide the fate of the target and pass that to the other. The virtual may have more degrees of effect, but those used by the constructive model are certainly among them. Only when the varying degrees of the outcome are significant to the objective do we need to force resolution into the virtual world. Resultant situations are far enough removed from these events that the outcome from either model will appear realistic. It seems that simple engagements can occur equally well in the constructive or virtual world with this simple unit.

Direct engagement between complex units is a bit different. The detection method has determined which unit or objects have been detected and by whom. Therefore, all possible targets are known to the respective weapon owners. If the constructive model calculates the outcome of the combat, it will result in a change in the overall unit strength of the complex target. The destruction of equipment will be identified down to the unique equipment types, at most. This means that the constructive model decides the number of each type that are destroyed, but not which individuals within that type. Transferring this information to the virtual model becomes difficult. Detection occurred between the constructive units, not between pieces of equipment. Therefore, it is possible for the constructive models to destroy equipment that was never detected by any object in the virtual world. Translating this type of event into the virtual world is difficult. Even were this not a problem, the constructive models do not identify the specific equipment that was destroyed. If the destroyed equipment must be mapped into the virtual world, it is possible to map the destruction onto the set of virtual detections (excluding the above problem). However, when the destruction information must be transferred to more than one virtual simulation the outcome of the arbitration in each may not be the same. This creates multiple versions of reality, although they are equivalent when constructed, they are not consistent with each other in the virtual world.

That last statement implies that, when direct engagement is performed in the virtual world, it will construct into an acceptable constructive solution. If fact the solution is more accurate than the constructive outcome. Direct engagement should, therefore, be resolved in the virtual world when a virtual unit is involved.

3.2.6 Engagement—Indirect

The second major form of engagement is indirect. To illustrate this, we will have to change our simple tank unit into a simple artillery unit composed of a single gun. When either the constructive or virtual version of the unit fires at a simple target, the result is the exchange of a single round from point A to point B.

The algorithms that calculate the damage may be different, but the basic outcome is the same, the death or damage of the simple target. Since detection is not necessary for indirect fire, there are no limitations caused by this process. The major impacts should come from the model's consideration of the effects of terrain and atmosphere on the trajectory, and the impact point of the shell. The virtual model may allow the terrain to diminish the effects of the round on the target due to screening effects, which protect the target from the explosion. The constructive model will probably give no consideration to this effect.

When a single artillery piece fires multiple rounds at a target, these may be dispersed by varying the bore sight of the gun or the amount of propellant used. The constructive model will take this into account when calculating the lethal area of fire. The virtual model will have to perform these changes and allow the algorithms to calculate the area of fire created. As a result, the area may be different, though equivalent. In the simple case, either the constructive or virtual model may be used with little loss of accuracy and with the infusion of little or no unrealistic effects.

For the complex case, a few new variables must be considered. A volley from an artillery battery is supported by multiple guns. When distributing the shots fired, the constructive model takes the rounds from a general pool of ammunition belonging to the entire unit. The virtual unit, on the other hand, wants to know how many rounds are fired by each gun. Since this type of information is determined by doctrine, a firing template can disaggregate the shot pool into specific, accurate shots by each virtual gun.

When the rounds impact the target area, the virtual dispersion pattern will be effected by the virtual placement of the artillery objects. However, since the patterns are controlled by the operators, this effect is minimized by human efforts to produce doctrinal lethal areas. In the constructive world, these areas are determined by templates and algorithms designed to recreate these effects. The major difference will be in the amount of variability present in the virtual placement as a result of equipment problems, fatigue, human mistakes, etc. The constructive guns will fire the same shot patterns again and again. Some of the virtual variability can be replicated in the constructive model by mistake generation algorithms.

The use of indirect fire can be calculated by either the constructive or virtual models, with little loss of accuracy. The fire comes like rain, falling on everything in the area. Except for extreme cases, such as an object hidden in a cave, the results will be acceptable from both models.

3.3 Timing

Virtual simulations expect to represent the events occurring in the world at a rate that is smooth to the human eye. This implies a time step of 1/15 to 1/30 of a second. Operating at these speeds, the simulation is able to provide realistic visual stimulation to the trainee.

Constructive simulations, on the other hand, typically interface with men who still experience much of the real world. Battalion commanders are used to operating with a combat situation that updates at discrete intervals, on the order of hours, or large fractions there of. The simulation, therefore, needs a time step which provides accurate event resolution to support this view of the battlefield. This has typically been set to one minute, although in some cases it is 15 minutes. Some method for integrating these two world views must be devised which satisfies both world needs.

One solution is to force all simulations to devolve to the lowest common denominator. This typically means synthesizing sub-time steps in the constructive simulation so that it operates on a virtual scale. Events are then broken into fifteen parts and released to the virtual world as virtual time passes. We have not added any fidelity to the constructive simulation in doing this.

The virtual simulation on the other side is sending update information to the constructive simulation 15 times a second. This is an added processing burden for the constructive, since it will not update its users more than once a minute. Neither can the added detail be used, since the constructive models are designed to operate on one minute events.

If, as we mentioned earlier, the virtual simulation is allowed to drive all interactive events, then what is needed by the constructive simulation is a filter. A gateway system can receive and process all of the virtual event updates. This information can then be released to the constructive simulation when it is needed, thus reducing unnecessary processing by the simulation. The gateway may convert constructive events into virtual events, and synthesize the necessary message packets going the other direction, as well.

Using this method, we create the appearance of both operating on similar time steps but have avoided modifying the existing simulations. The cost of this is the addition of another module to serve as a gateway. There will be some dissimilarities between the constructive and virtual worlds, but these should be visible only to those who have access to both. Since the training audience does not have this access, they will not perceive these artifacts. One example is the event in which a virtual unit has zig-zagged its way between point A and B. The constructive simulation represents this as a straight line move. When constructive artillery is fired at some point along the straight line, the event is resolved in the constructive simulation and a hit occurs. If this event is resolved in the virtual, a miss will occur. Explaining the miss to the constructive trainee is difficult, since he has no other means of hitting such a target. This artifact can be exploited by virtual players if they are aware that they are in such a situation. But, gamesmanship of this type not uncommon today, and is not a result of vertical integration.

3.4 Disaggregation

In exploring vertical integration, we have already touched on the idea of disaggregation of constructive units and the aggregation of virtual units. We will address these ideas more fully in the following two sections.

Disaggregation is typically applied to complex constructive units. The purpose is to transform these into acceptably accurate virtual objects. This includes the object's location, orientation, physical condition, and combat history. This is one of the primary problems that was addressed in the vertical integration prototypes described in chapter two.

3.4.1 Units

Beginning with a complex unit occupying a single location, our goal is to produce realistic locations and orientations for all of the individual objects that make up that unit. This information will then be passed to the virtual simulations for their operations.

Disaggregation can be divided into three layers of operation (figure 3-5). The first layer of this transformation has typically been a doctrinal template. This technique has been used for more than ten years to create specific object locations for the TACSIM

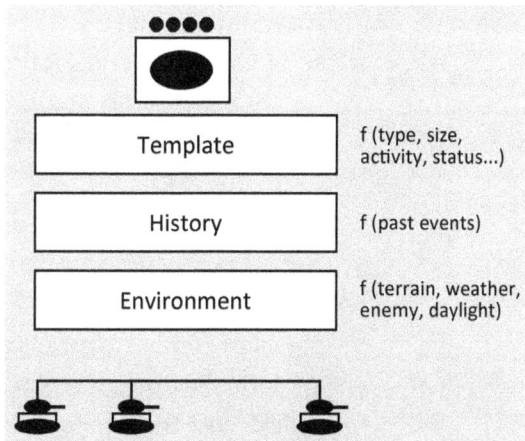

Figure 3-5. Layers of Disaggregation

simulation. More recently, it has been applied to create an interface between CBS and TACSIM. Figure 3-6 shows the unit location provided by CBS, with a military icon. The locations of all of the objects templated by TACSIM are given by the smaller symbols.

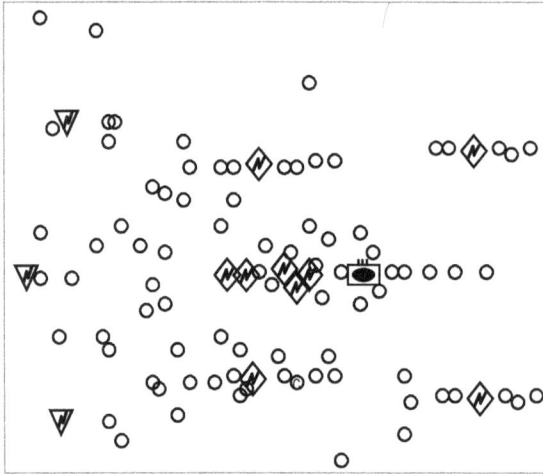

Figure 3-6. CBS-TACSIM Disaggregation Template

Since military units have a doctrine which describes their layout based on the location of the command post, template information can be lifted directly from the manuals. This is often modified by the most current experience of military experts. Imagine the command post as the trunk of a tree. The objects are then laid out around it in a branching fashion according to sub-organizations. A Battalion may contain 3 companies, each of these may contain 3 platoons, and each of these may contain 3 squads. All of these operate is specific patterns based upon the location of the Battalion headquarters. Templates vary according to the constructive unit's type, size, strength, activity, and status.

When the objects are laid out for the first time, they have not yet experienced any individual events. Each object type, or sub-unit, is basically identical. Each should have the same combat experience, damage, fuel ratio, and supply ratio. But, once a unit has been disaggregated, they quickly lose their uniformity. One may have burned more fuel, another may have experienced significant damage, a third destroyed. These differences must be maintained from disaggregation to disaggregation. If this information is lost in re-aggregation, much of the value of the virtual interface is lost. The virtual trainees can no longer see cumulative effects of their actions. They are fighting a force with magical powers of reconstitution. Once a unit has experienced disaggregation, its entity state must be saved for later use, even while it is aggregated. These unique characteristics are the second layer of disaggregation.

The third layer of disaggregation is the environment. This includes the terrain, weather, daylight, combat conditions, and location of the enemy. This layer is the most complex and time, consuming to calculate. Terrain modifies object placement by introducing limitations on its performance. Obvious examples are: placing objects in lakes, dense forests, and on mountain tops. Deployment patterns must be modified to adapt to these in a realistic manner. One of the guiding factors here is maintaining line-of-sight between objects of the same unit. This provides a strong force in moving the objects. Weather changes the operational environment. Muddy roads, fog, and rainfall may cause the objects to be deployed closer together to provide the visibility desired between objects and to reflect movement problems. Units operate differently during dusk and at night, than during the day.

Combat conditions include the strength of the unit and the chemical or nuclear environment. If is half-destroyed, it is probably operating in a more defensive posture. In chemical or nuclear contamination areas, the operation of the unit is much more cautious and may dictate defensive movement most of the time. Proximity and experience with the enemy also impacts the placement of objects.

Units that have recently come under sniper or artillery fire are much more defensive and cautious in their movement. Those that know the locations of the enemy are also operate in a different mode.

3.4.2 Ordnance

Disaggregating ordnance is necessary when it is remote, or independent, from the firing unit. Examples of this are: artillery fire, aircraft bombs and missiles, and minefields. These weapons depart from the originating unit and take on very distinct characteristics, different from those of the unit that fired them.

When an artillery piece fires a volley of rounds, these are placed in a very determined pattern around the targeted location. Given that an artillery unit is constructive, the fired volley can be disaggregated according to the desired destructive effect, as described earlier. It is not necessary to disaggregate the unit in order to determine this information. The engagement is actually between the volley of rounds and the target, with little consideration for the firing unit. Since fire patterns are doctrinally taught, disaggregating these can be done similarly to disaggregating a unit. A doctrinal template can be used to place each round initially. These locations can then be adjusted by weather and terrain conditions. As a result, it is possible to conduct constructive indirect fire against virtual targets by disaggregating the volley, but not the firing unit.

Should extremely accurate virtual patterns be necessary, there is no reason the constructive unit, itself, could not be disaggregated before initiating the fire order. The fire mission order would then effectively be several orders given to each gun in the battery.

If the problem is reversed, virtual artillery firing at a constructive target, the situation is less difficult. When a virtual artillery weapon hits a constructive target, the effects should be the same as the effect of a single constructive artillery round. Multiple virtual rounds can produce multiple engagements. These rounds may also be buffered and delivered to the constructive target in a single constructed volley. Either way the effects will be acceptable in the constructive world.

The effect is very similar when bombs and missiles are delivered from aircraft. We must qualify these weapons as either smart or dumb. Smart weapons may be guided by some sensor, while dumb bombs follow a descent trajectory determined by the physics of the weapon falling through the air. Dumb bombs are delivered by aircraft weapons systems designed to disperse them into a given pattern, similar to the artillery shells. These can most definitely be modeled with the same techniques used above.

Smart bombs may be considered to have a similar type of decision making process as a manned unit. These take direction from a sensor, adjusting their trajectory according to update information. This can be modeled specifically, or the effect can be included in an algorithm which replicates it to an acceptable level of detail. In either case these weapons become virtual entities to produce virtual effects.

The effects of human-emplaced minefields can be templated, since they are doctrinally laid. Air delivered minefields are randomly laid according to a dumb bomb trajectory pattern; these may also be templated. Placing each mine in the virtual world carries some of the overhead of disaggregating a unit. A history should be maintained so that each disaggregation produces the same pattern. This creates a situation which virtual trainees can reason about, and react to, in a natural way. Should a man follow in the tread marks of a tank that passed earlier, he should also pass through the minefield unharmed. It would be incorrect for him to be destroyed as if the minefield had been newly laid using different dispersion patterns, when, in fact, it had not.

Directed energy weapons include lasers, radio frequencies, electromagnetic pulse, and sound. All but the laser are area weapons, designed to blanket a limited area with their effects. These effects are essentially the same as the artillery volley, but they are more uniform and continuous within the area. They can be modelled with the same methods.

Lasers can be area or pinpoint weapons. The area can be several meters across and still be effective. These may be modeled more like direct-fire weapons. They have little variation once fired and proceed to the target in a predictable pattern. It is more accurate to create a virtual representation of them. This may be a fan beam at the impact point. These weapons are used to burn solid objects, particularly fuel inside of storage containers. Pinpoint lasers are directed at a very specific target, and are best modeled in the virtual world. The exactness of the weapons, effective at burning electro-optical and biological sensors, makes them unwieldy to model in the constructive world.

Lasers may also be used as target designators. In this case, they are actually part of the sensor system on a smart weapon. These are affected by terrain and environment, and belong in the sensor model.

3.4.3 Consumption

All armies at war have an enormous appetite for supplies.
These are most dominantly fuel, ammunition, and food,
but may also include medical supplies, tools, mechanical
parts, and toilet paper. A constructive simulation meets
these needs by providing supplies to keep units moving
and fighting. This usually consists of one type of fuel for
all vehicles and food by the pound. Ammunition can be
handled as a single type or as a number of different types.
In the latter case, the supply units must deliver the appropriate amount of tank, artillery, and small arms rounds.
These are then distributed evenly among the weapons
systems that can use them. The algorithms are usually
designed to provide the amount of fuel and ammunition
that will keep all of the equipment in the unit fighting for
the same period of time. This minimizes the number of
supply convoys needed to service the unit, since special
cases do not arise for one weapon. Ammunition may also
be measured in pounds. In this case, the unit will have an
algorithm that burns the supplies at a given rate, which
does not count individual shots. In either case, keeping
the entire unit operating for the same period of time is the
primary goal.

When a unit disaggregates, each object begins to burn
supplies at a different rate, using the virtual algorithms.
This is one of the characteristics of individuality that we
attempted to retain in a previous section. Unfortunately,
the goal of keeping supplies divided so that the entire
unit can operate for the same period of time is now very
difficult. It is unlikely that any algorithm will be able to
determine the constant burn-rate of each vehicle type, in
an efficient manner, in order to distribute the supplies.

Dividing each type of supply by the number of objects and giving each an equal share is not appropriate. A jeep definitely requires less fuel than a tank. In fact, an equal share of the fuel may exceed the maximum capacity of the jeep.

The most efficient and equitable solution is to divide the supplies into weighted shares. Each object receives a portion equivalent to its maximum capacity. This will not guarantee that each object will be operational for the same period of time, but it is a best estimate. If each virtual object had a parameter that described its burn-rate in terms of a common unit of measure, this could be used to distribute the supplies. This would create the most equitable solution.

Although we wish to disaggregate in the most realistic fashion, even an exact algorithm can not insure that the objects will operate for the same amount of time. The virtual model will be consuming supplies for each object, independently. If the unit were simply a convoy moving down a road, the rates may remain about the same. But once it engages in any type of unique behavior, that relationship will be lost. This characteristic must be maintained even after the unit has been re-aggregated.

It is possible that even the disaggregated objects are being resupplied by the constructive model. In this case, the supplies do not go to each individual object, but to a supply point. The objects are then responsible for going to that point to get what they need. For the constructive supply algorithm to know what is needed, the virtual objects must periodically update the supply parameters of the constructive unit. This will inform the constructive model of the time and amount of supplies that are needed.

The operations of the virtual supply point are a separate model from the supply convoy generation model. The first will pour supplies into the supply point by the constructive convoy. After that, all distribution can be handled by the virtual model. However, the constructive model needs to account for the supplies in the objects and at the supply point. The model should not be asked to provide additional supplies simply because the objects are almost empty and the supply point is full.

Reaggregation of supplies is much easier. All supplies become the property of the single unit, which by definition, will then be able to operate together for a given amount of time.

3.4.4 Orders

Orders to perform certain missions and actions typically begin at the top of the command chain and work their way downward. Each commander adds detail along the way. In order to simulate these people, this disaggregation of the order must be simulated. The problem of disaggregating an order is one of the most complex. When a human performs this function, the basis is the doctrine taught to him. This is certainly modified according to the unit's situation. But, even the basis for these modifications is set in military doctrine. Once a doctrinal order template has been applied, the specific situation will modify this. The modifications include the terrain, unit type, unit echelon, enemy location, unit situation, and time of day. Although many rules may be used to perform these, Monte Carlo functions can also be used to provide variations in these effects.

It is possible for the situation, particularly with respect to the enemy location, to nullify or seriously modify the intended order. For example, the order to move forward may be impeded by terrain. In this case, goal programming would lead to a path which maximizes compliance with the order, while minimizing terrain effects (Figure 3-7). This will result in a path around the side of the hill, rather than over its peak.

Objective Force Curve

Hill

Move Path

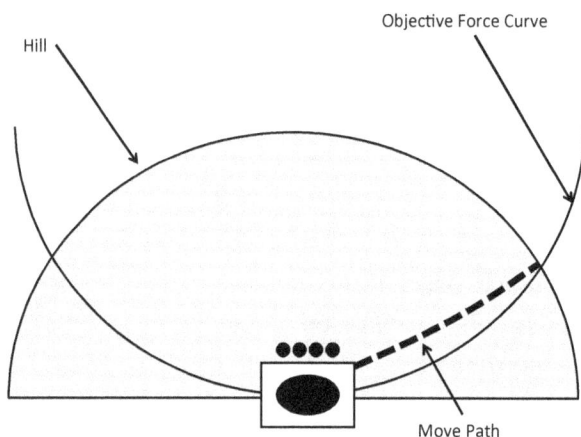

Figure 3-7. Interaction of Objective and Environment

Should the hills be promoted to an impassable mountain range, the order to move forward may be totally nullified. Assuming that unit situations are being reported back to the human operator, this situation may need to be addressed manually. Certainly, many different levels of problem resolution may be programmed into the simulation. However, there is always a limit in the amount of processing that can be expended on a problem.

3.4.5 Firewalls

The rules for vertical integration may cause some very unwelcome disaggregation results. Since many of the actions are resolved by operating at the virtual level, a domino effect may occur which causes a much larger number of units to disaggregate when we are trying to resolve a single event. To prevent this, we must install some firewall rules which limit the amount of disaggregation that can be generated by any single action.

One method is to disaggregate only those constructive units that are directly in contact with the virtual object stimulating the disaggregation. This implies that the disaggregation of a unit for virtual combat cannot cause another neighboring constructive unit to disaggregate in order to join in the fight. Should the second neighbor rightly belong in the fight, that will be computed at the constructive level. The results will then trickle down into the virtual world during later engagements. This method may result in unrealistic combat sets. Calculating these separately in the constructive models, is an attempt to rectify this problem. Although the best solution is to resolve all conflict at the virtual level, realism must be tempered by what can be done with the available technology.

Another method is to disaggregate a constructive unit only inside the virtual unit's node. To all external simulation nodes, the constructive unit remains constructive. This is very useful for sensor detection events of individual objects all across the battlefield. This detection characteristic need not trigger a constructive unit into disaggregation when the sensor sweeps over it. Instead, all of the disaggregation is performed only inside of the sensor model. This allows it to detect objects without causing any disaggregation events. This firewall is particularly useful here, since each sensor is independent of the others, and since they can have such a pervasive effect on the constructive world if not controlled.

It may be useful to set disaggregation horizons for different types of events. These horizons would limit the

amount of disaggregation that can be performed. This is
a more flexible version of the first firewall. The horizon
may be defined by a range from the causal object, or it
may be defined as a number of objects instantiated in the
virtual world. The latter measurement is driven by the fact
that any virtual unit has a limited amount of attention
to give to objects. Encountering 1000 objects or 10,000
objects may result in the exact same behavior. There is a
point beyond which the addition of new objects does not
change the course of action. If a human is visually observ-
ing the scene, the number of objects that can be displayed
is certainly limited to the number of pixels on the display.
A more practical limit is determined by the boundaries of
human perception and the ability to absorb information.
This type of firewall is most useful in densely populated
areas where the action is the hottest.

When a virtual unit encounters a constructive unit, we
have assumed that disaggregation is performed on the entire unit. In reality, the virtual object will only encounter
a portion of the constructive unit. If we could determine
the specific portion, we could disaggregate only that part
of the constructive unit. This limitation will reduce the
domino effect, since it limits the initial cause.

Finally, we may begin the simulation with a list of units
which are designated by a human controller as being
allowed to disaggregate. Only these may perform that
function, all others must remain in their natural condition. This will cause problems in adhering to earlier
established methods for dealing with virtual-constructive
events. It implies that methods must be available for
dealing with every combination of virtual-constructive
event resolution.

3.5 Aggregation

The process of aggregating virtual information into constructive information is considerably easier than disaggregation. Since it removes, rather than adds, information. We need a scheme that is consistent and that takes into account the types of events that can occur later.

In earlier sections, we described the need to maintain disaggregation characteristics, even after the item had been aggregated again. Although we may be removing information from the active part of the item, these unique characteristics must be stored in a manner that allows us to apply them to the next disaggregation.

3.5.1 Units

To constructive units, we are primarily creating a new location and identifier for them. The identifier in military simulations is typically a type (armor, artillery, supply convoy), echelon (company, battalion, brigade), and name (2nd Armored Division, 511 Military Intelligence Battalion). Virtual objects contain information which indicate their parentage, which can be used to generate the above descriptors for the constructed unit.

Generating a location can be done in several different ways. One is to average the locations of the all of the objects being aggregated. This results in a center of mass of the unit. Although geographically descriptive, this does not describe the new unit in a militarily accurate manner. The location of a company is usually given as the location of the unit command post and a radius inside of which all of its objects can be found. To create this type of ag-

gregation, we need only know which of the objects being
aggregated is the command post. Aggregation of location
then becomes a much easier problem. Since the command
post controls the type, echelon, and name of the unit, the
objects being aggregated can come from many different
command structures. This flexibility allows us to create
units from the objects of any number of other units.

3.5.2 Ordinance

In aggregating ordinance, we are referring to fired, virtual, indirect-fire weapons which are about to impact a
constructive target. Since a constructive unit is a single
location with a radius of dispersion, the virtual weapon
can be handled exactly like the firing of a single round of
constructive ordinance at the target, assuming that we
want to determine the outcome of the event in the constructive world. In earlier sections, we indicated our desire
to do these calculations in the virtual world. We mention
the alternative here to show that it is possible to handle
the events in either way. In constructing the ordinance,
you lose much of the precision targeting that may have
been performed in the virtual world. Although a hit on
a unit may be assured, the target may have been a very
specific piece of equipment. The constructive solution will
probably not destroy that exact piece of equipment. Since
this detail is not being maintained, it is assumed that this
method is being used only when the outcome is not being
directly observed by the virtual training audience. This
may best be used for indirect-fire engagements. Direct-
fire will cause several inconsistencies, due to the scene
presented to the virtual simulation operator.

3.5.3 Consumption

The aggregation of consumables is a very basic problem. Given that the constructive template describes the different categories of these, virtual representations of these must be summed into these categories. Assuming that the disaggregation derived all of the virtual types from earlier constructive types, virtual consumables can be returned to the constructive types from which they came.

3.5.4 Orders

Since orders result in objects having an objective, we shall refer to this as aggregating the objective of the objects into an objective for an entire unit. For this objective to exist at the virtual level, it must have passed through several layers of constructive military units, be they real or simulated. An earlier constructive version of the objective may be used as that of the newly constructed unit. When this is not available, we may collect the objectives of the objects and build from them a set of unit objectives. Some parts of this are the same as the location problem above, but rather than seeking the unit's current location, we are seeking its intended future location. This can be averaged from the objects or taken directly from that of the command post object. Characteristics, such as the amount of ordinance to use and acceptable casualties, can be taken from the objects in a similar manner.

Organizational Architectures

1 1 0 0 1 0 0 0 1 1 0 0 1 0 1 0 0 0 1 1 0 0 1 0 1 0 1 0 1 0 0 0 1 1 0 0 1 0 1 1 1 0 0 1 1 1 0 0 1 0 1 0

The purpose of this section is to determine the best organization of the simulation nodes and the connections between them. In both the constructive and virtual worlds, the simulations are typically connected to a common backbone in order to exchange information using ALSP or DIS. "Backbone" refers to the structure where each simulator shares the same network line with other simulators, similar to the old party-lines of the telephone system. Every node receives, and has access to, every message of the network. It is the responsibility of the node to determine which of these messages are intended for or useful to it. The connection between the constructive and virtual worlds will use a set of gateways to bridge the gap. In order to minimize the amount of network traffic and communication delays, these gateways should be organized in a more structured fashion. Figure 1-4 depicts a single "Interface" between the two worlds. This merely illustrates the need to bridge the gap. Figure 1-6 describes some more specific methods for doing this. These are:

- Model Level
- Military Unit Level
- Military Command Level
- Functional Level
- Geographic Areas

4.1 Network Architecture

The DIS network is currently structured as a single back-bone, using broadcast communications. The methods listed above require that this structure be broken into more efficient pieces.

4.1.1 Backbone

The network backbone is a simple structure that is very dynamic. Any node can join or leave by simply broadcasting a single message. It does not have to register its existence with a network server. Broadcast communication is also simple. A single message on the network is delivered to every node. It is then up to the nodes to determine whether it is useful to them. But each node impacts all of the others. The backbone is a bottleneck for communications. It limits the amount of growth possible since no operations can be isolated and contained in a smaller area. Two nodes representing adjacent military objects may exchange information very frequently. These messages are delivered to every other node on the backbone as well. The traffic between two closely coupled simulators is mixed in with all traffic, adding to the network burden.

4.1.2 Internet

A system that is able to grow to host 10,000 or 100,000 individual objects must be able to partition the network into manageable pieces. Closely coupled units, like those mentioned above, should be grouped and isolated. Only information affecting a larger domain should escape this isolation. The structure we are looking for is that used by the Internet. A set of parent and child nodes, with no real top or bottom.

Closely coupled nodes are placed on a LAN with a gateway to a wider network domain. Each level is also on a MAN or WAN of limited scope. The entire structure is similar to the human central nervous system. But, there is one important difference, although each piece may have specialized functions, there is not necessarily a head. The Internet was originally designed to provide a dynamic, survivable command and control structure for the military in the event of severe disasters, such as nuclear war. It is intended to have no single point of failure and no single communication path between two nodes. This design has also made it very scaleable. It has grown to several million nodes and its growth is still not limited by its basic architecture (although there are other implementation limits).

The DIS network needs to be restructured into something similar. Although this adds complexity, it is necessary to accommodate the incredible growth envisioned. Once this is done, designers will be free to explore some of the simulation node organizations described in the next section.

4.2 Organizational Alternatives

There are many ways to organize the members of a simulation network beyond the simple broadcast backbone. Adding structure adds complexity, but it also increases efficiency. This efficiency is essential to allow what has been a very successful laboratory experiment to grow into the envisioned operational system.

4.2.1 Model Level

The most basic approach is to assign a gateway process to each of the constructive models and add it to the virtual network backbone (Figure 4-1). This will result in every message on the network being processed by each of the constructive gateways. Information filtering is then carried out inside each one of these to provide the data needed in the constructive or virtual worlds. Data flowing from the virtual to the constructive worlds is filtered to provide event time steps similar to the constructive model being served. Data flowing the other direction is enhanced to serve the same function. Filters pass only information on events and units that are of interest to the receiving simulation. If there is only one model processing underwater events, then the status of entities such as submarines may not be transmitted through the gateways. But, if there are even two models processing this information, it must be placed on the network. This will cause additional network traffic and processing load for every simulation.

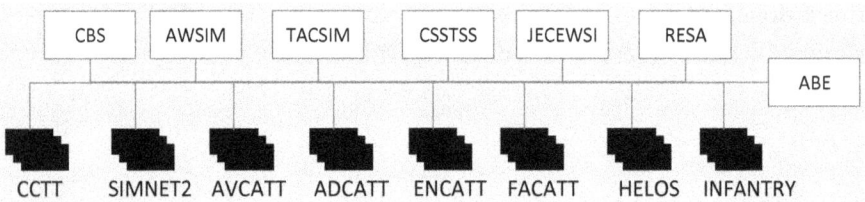

Figure 4-1. Model Level

A modification to this method is to use a single gateway to the constructive world. This performs filtering based on the union of information needed in that world. After initial processing, information is passed to another gateway used by each of the constructive simulations. Although this reduces the amount of duplicate processing being performed, it creates a bottleneck between the two worlds. The flow of data intended for any one of the models will impact that intended for all of the others. This also requires a degree of cooperation and compromise that can be avoided by giving each constructive model its own gateway.

An advantage to the second method is that it allows the constructive simulations to send data to each other, without impacting the virtual nodes on the DIS backbone. Many units in the envisioned exercises will exist only in the constructive world. Information about these is necessary only for the other constructive models. This method for limiting the broadcast of this information can be very valuable.

4.2.2 Military Unit Level

The second method is to provide a gateway for each constructive unit (Figure 4-2). The DIS units of an armor company and the constructive representation of that company

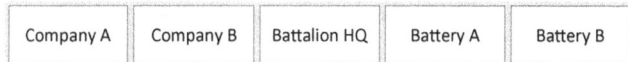

| Company A | Company B | Battalion HQ | Battery A | Battery B |

Figure 4-2. Military Unit Level

would form a LAN. This would then have its own gateway to the rest of the simulation network. Information about the unit can then be shared locally, without disturbing the entire simulation world.

This creates network nodes and LANs that represent military units. Units can then be added or removed very cleanly. In the purest form, this requires that a copy of each constructive model be used for each military unit. A single copy of CBS would no longer control the entire ground combat. Instead, many CBS models would be running, each responsible for a single unit, and ghosting all of the others.

For practical purposes (such as a limited number of host computers), multiple units may be controlled by the same copy of CBS, but they would be connected virtually, as if they were different models. The disadvantage of this is that there remains a backbone, which is a bottleneck for communications between units.

4.2.3 Military Command Level

The models can be built and organized as the military structure (Figure 4-3). Although the network is actually a flat hierarchy, the relationships of the models form a virtual command structure within this. A separate model is run for each level of command. These communicate up and down according to the type of information that is usually passed between military units. Each level contains the decision models required to operate at that level. These can then be joined together in a natural manner, since the communications are a standard set of known information (see Chapter 3).

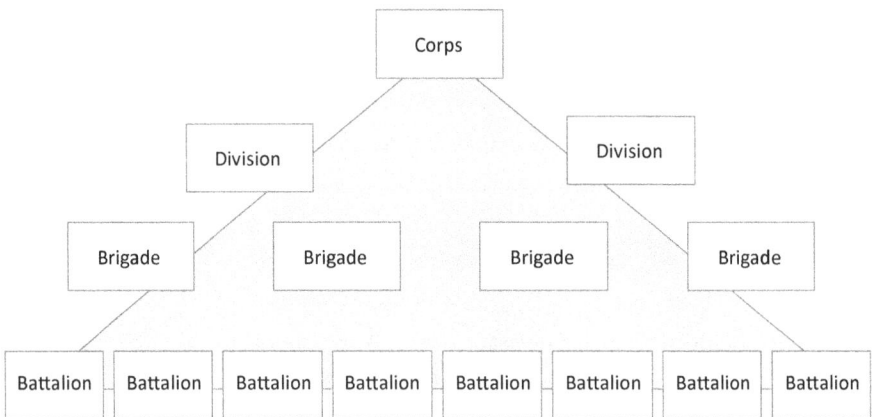

Figure 4-3. Military Command Level

This organizes models according to the objects being simulated rather than placing the all of the objects together within the model.

4.2.4 Functional Level

Division at the functional level results in creating local groups of entities that perform the same function (Figure 4-4). For example, all units that move would be organized into a group with a gateway. This would also be done for units that

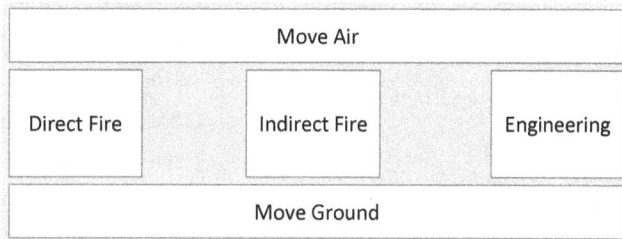

Figure 4-4. Operational Level

perform direct-fire, indirect-fire, resupply, sensor collection, air-to-surface engagement, surface-to-air engagement, etc.

It is obvious that each unit will belong to several virtual sub-networks. One advantage to this method is that a single copy of each model is operating, rather than multiple copies. This is the technique attempted with the CBS model. The CBS simulation engine is a single process. Experiments have been conducted in order to split this into multiple processes, which can then be run on multiple processors. Due to the functional structure of the model, these attempts have been along functional lines. The developers have tried to isolate modules, such as the attrition calculations, and make them a separate process. The functional divisions are successful, but the latency caused by communicating with a separate process has resulted in unacceptable operational results. Although the time required to query a separate process is a fraction of a second, the fact that thousands of these are required every game-minute has resulted in a model that can not maintain a real time processing rate.

4.2.5 Geographic Areas

The final method is to divide the battlefield into several geographic areas. A copy of each constructive model is then run for each of these areas (Figure 4-5). Each area is then assigned a single gateway which is used when events cross geographic boundaries.

52N 10E	52N 11E	52N 12E
51N 10E	51N 11E	51N 12E
50N 10E	50N 11E	50N 12E

Figure 4-5. Geographic Areas

Units within this area can then fully operate with each other, without impacting any of the other nodes on the network. Only when an event, such as artillery fire, crosses these geographic boundaries is a message generated.

The parallel processing community has found this method of division to be the most efficient. Experience has led to some insight into the best ways to distribute functions. They have found, as illustrated in chapter 2, that the best division of processes is across geographic, rather than functional, boundaries. The interactions between entities are much more frequent among those in the same area, and less frequent with those more distant. This type of division minimizes processor node communication by processing geographically-local events on the same processor.

4.3 Computer Generated Forces

Computer Generated Forces (CGF), also known as Semi-Automated Forces (SAFOR or SAF), are very popular in the virtual community right now.

4.3.1 Virtual

As very excellent vehicle simulators developed, the need for a reactive opponent grew increasingly important. Unfortunately, there were insufficient humans or computers to operate the vast numbers of forces found on the battlefield. To fill this gap, they created SAFOR systems, which typically involves a human operator providing basic input to dozens or hundreds of objects, and an expert system adding the specific details to be displayed. For example, the operator may specify that a tank move from point A to point B. The expert system and algorithms would then adjust this straight-line path to account for the terrain and obstacles encountered, generating a more realistic travel path. Refer to chapter 2 for more details.

4.3.2 Constructive

In many ways, CGF is a visual version of what constructive models have been doing since World War II. Basic operations are required from a human, and the rest is supplied by the computer. Even board games automated outcome results by turning the work over to a set of dice and decision tables.

Constructive simulations have usually served a different type of customer. The force commanders entered the basic unit movements and actions—*"move to position A, attack*

unit K, array for defense, etc." The computer then decided
how to control the specific actions and events that occur
under the conditions provided. The emphasis was not
on the visual scene, but on the outcome of events at the
military unit level.

4.3.3 Future

Both approaches have similar goals. In order to join the
two levels, a certain degree of automation will have to be
provided. This will not focus on the visual scene or the
event outcomes. It will focus on the command and control
jobs performed by thousands of intermediate soldiers, between the highest levels of command and the individual
combat vehicles.

The models will represent the cognitive decision making
process of human minds. One part of this will be the
doctrine drilled into each military mind in basic and
advanced training. Another part must come from fields
like artificial intelligence, expert systems, and cognitive
reasoning. The simulation must replicate the decisions of
privates carrying rifles, sergeants leading platoons, staff
officers providing advice, and command officers making
decisions. The scale of reaction times for each of these dif-
fers from seconds to hours.

Research into this type of model is in its infancy. ARPA
has begun with four projects which will be applicable to
the virtual connections we are exploring. The Combat
Instruction Sets (CIS) attempt to formally describe the
types of events that take place on the battlefield. The
SAFOR projects then use these relationships to supply
templates to human operators. They use them to acceler-

ate the speed at which they can control combat entities.
The next step is to create Computerized Forces (CFOR)
which will be able to select from CISs in response to any
situation encountered. Finally, the Intelligent Forces
(IFOR) program seeks to create models which can create
and modify CIS as needed to respond to the enemy. IFOR
is very ambitious and has attracted many critics.

4.4 Prototype Organizations

In Chapter 2, we described several prototype projects
which are being pursued in vertical integration. The
primary characteristic of these projects is that they create
a 1-on-1 situation between a constructive and a virtual
model. They need to be extended to include M-on-N
simulations and simulators.

When Eagle passes units into the SAFOR world, it freezes
the constructive unit in place while the virtual world
controls it. The units need to continue to exist constructively and to operate as aggregated cases of their virtual
components. The virtual objects need to be able to operate in an automated fashion without human intervention.
The vertical bridge between the simulations needs to be
invisible so that the training audience does not know or
care that the events are being handled elsewhere.

These types of improvements are necessary to create a
simulation that can be used to train soldiers. This will
require an M-on-N capability.

4.5 Preferred Organization

4.5.1 Pure Geographic

Of all of the structures discussed, one is clearly prefer-
able—geographic areas. This has the advantage of isolat-
ing effects and grouping them under a single processor
or process. The parallel processing community has done
extensive work to illustrate that logical processes with
geographic boundaries is the most efficient structure.

This structure also has the ability to be pyramided into
larger sections. This gives us the internet-like qualities
of locality and independence. The geographic area to be
assigned to each processor must be determined with care.
It must not be too small, or the strengths of the structure
will not be realized, i.e., objects/units will continually be
communicating across boundaries. Conversely the areas
must not be too large, or each processor will be over-
loaded, and hence slowed, by the amount of calculation
required for each one.

4.5.2 Geographic/Command Hybrid

An improved solution may be a hybrid. Underlying the
geographic structure, we may include a command structure. This will allow us to assign a separate process to each
object and still group these into a geographic LAN. Since
a military unit communicates most heavily with other
members of its unit, this underlying structure will isolate
much traffic from other objects/processes in the same
geographic area.

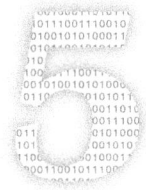

Recommendation for Future Work

In this work, we have explored methods for integrating constructive and virtual level simulations. The military has a growing need to create a single integrated environment, where the strengths of both types of simulation are used to enhance the training experience generated by computers. The virtual simulation community is beginning to develop semi-automated forces (SAFOR) algorithms which will allow them to join with constructive simulations. Enhanced versions of SAFOR will be the bridge that connects the constructive and virtual worlds. These will perform the fidelity enhancement and removal needed for two-way communications across the bridge.

The simple broadcast backbone networks used in both the constructive and virtual communities will not support the types of exercises envisioned for the military. A more complex structure must be imposed on the network, something similar to the Internet. The nodes will simulate events in specified geographic areas. Within a node or LAN, these areas may be further subdivided and organized into a military command structure.

Research is needed to identify the sizes of the geographic areas to be represented in each simulation node or LAN. These sizes may vary by the type of exercise being conducted and by the type of activity in each. Dynamic, heterogeneous areas are very desirable. These will allow the simulation to be structured most efficiently, and give it the flexibility to adjust itself as simulated events change the activities in the defined geographic areas.

Although SAFOR are very useful, the development of Command Forces (CFOR) and Intelligent Forces (IFOR) is essential to create a fully automated environment that contains all of the realism of a full man-in-the-loop environment.

Bibliography

1 1 0 0 1 0 0 0 1 1 0 0 1 0 1 0 0 0 1 1 0 0 1 0 1 0 1 0 1 0 0 0 1 1 0 0 1 0 1 1 1 0 0 1 1 1 0 0 1 0 1 0

"Aggregate Level Simulation Protocol Operational Specification", MITRE Informal Report, August 1993.

"Aggregate Level Simulation Protocol: Tactical Simulation Model Interface Control Document", MITRE Technical Report, July 15, 1992.

Allen, Gary W. and Smith, Roger D., "After Action Review in Military Training Simulation", *Proceedings of the Winter Simulation Conference*, December, 1994.

Allen, Gary W. and Smith, Roger D., "After Action Review in Military Intelligence Simulation", *Georgia Tech Annual Course on Modeling, Simulation, and Gaming of Warfare*, September, 1994.

Allen, Gary W. and Smith, Roger D., "The Tactical Simulation: The Army's Leading Intelligence Collection and Dissemination Model", *Phalanx*, March-April 1994.

Allen, Gary W. and Smith, Roger D., "Immediate Feedback on Training Performance", *Proceedings of Orlando Multimedia'94*, February, 1994.

Allen, Thomas B., *War Games*, Berkley Books, New York, NY, 1987.
Ammar, Hany H. and Deng, Su, "ACM Transactions on Modeling and Computer Simulation", *ACM Transactions on Modeling and Computer Simulation*, Volume 2 Number 2, New York, NY, April, 1992.

Bagrodia, Rajive L., Chandy, K. Mani, Misra, Jayadev, "A Message-Based Approach to Discrete-Event Simulation", *IEEE Transactions on Software Engineering*, Vol. SE-13, No. 6, June, 1987.

Bates, Joseph, "The Nature of Characters in Interactive Worlds and The Oz Project", Technical Report CMU-CS-92-200, Carnegie Mellon University, Pittsburgh, PA, 1992.

Booker, Lashon, "Command Entity Architecture", Presentation of the Advanced Research Projects Agency, 1994.

Brann, J. Joseph, "WARSIM Time Synchronization/Management", Unpublished, Loral Federal Systems Company, June, 1994.

Brustoloni, Jose C. and Bershad, Brian N.,"Simple Protocol Processing for High-Bandwidth Low-Latency Networking", Technical Report CMU-CS-93-132, Carnegie Mellon University, March 1992.

Butler, Ralph and Lusk, Ewing, *User's Guide to the p4 Parallel Programming System*, Argonne National Laboratory, Argonne, Illinois, 1992.

C4I for the Warrior, Joint Chiefs of Staff, U.S. Department of Defense, 4 September, 1992.

Caceres, Ramon, "Measurements of Wide Area Internet Traffic", Progress Report No. 89.12, University of California, Berkeley, CA, December, 1989.

Chandy, K.M. and Sherman R., "Space, time, and simulation", *Proceedings of the SCS Multiconference on Distributed Simulations, Society for Computer Simulation*, San Diego, CA, 1989.

Chandy, K. Mani and Misra, Jayadev, "Distributed Simulation: A Case Study in Design and Verification of Distributed Programs", *IEEE Transactions on Software Engineering*, Vol. SE-5, No. 5, September, 1979.

Chany, K. Mani, Holmes, Victor, and Misra, J., "Distributed Simulation of Networks", *Computer Networks 3*, North-Holland Publishing Company, 1979.

"Close Combat Tactical Trainer—Training Concept", Unpublished Paper, Dynamics Research Corporation, Wilmington, MA, 1992.

Comer, Douglas E. and Stevens, David L., *Internetworking with TCP/IP Volume III: Client-Server Programming and Applications*, Prentice Hall, Englewood Cliffs, NJ, 1993.

Communication Architecture for Distributed Interactive Simulation, Institute for Simulation and Training, Orlando, Florida, 15 May 1992.

"Computer Generated Forces Standards Rationale", DIS-CGF Mailing List, Orlando, Florida, April 1994.

Corps Battle Simulation: Analyst's Guide—Air, Jet Propulsion Laboratory, Pasadena, California, 1993.

Corps Battle Simulation: Analyst's Guide—Ground, Jet Propulsion Laboratory, Pasadena, California, 1993.

Corps Battle Simulation: Analyst's Guide—Logistics, Jet Propulsion Laboratory, Pasadena, California, 1993.

Corps Battle Simulation: CBS-AWSIM Interface Control Document, Jet Propulsion Laboratory, Pasadena, California, 1993.

Corps Battle Simulation: CBS-CSSTSS Interface Control Document, Jet Propulsion Laboratory, Pasadena, California, 1993.

Corps Battle Simulation: CBS-TACSIM Interface Control Document, Jet Propulsion Laboratory, Pasadena, California, 1993.

Corps Battle Simulation: COAST User's Guide, Jet Propulsion Laboratory, Pasadena, California, 1993.

Corps Battle Simulation: COBRA User's Guide, Jet Propulsion Laboratory, Pasadena, California, 1993.

Corps Battle Simulation: Database Description Document, Jet Propulsion Laboratory, Pasadena, California, 1993.

Corps Battle Simulation: Executive Overview, Jet Propulsion Laboratory, Pasadena, California, 1993.

Corps Battle Simulation: Master Interface Control Document with Addendum, Jet Propulsion Laboratory, Pasadena, California, 1993.

Corps Battle Simulation: Operator's Manual, Jet Propulsion Laboratory, Pasadena, California, 1993.

Corps Battle Simulation: Order Screens and Help Document, Jet Propulsion Laboratory, Pasadena, California, 1993.

Corps Battle Simulation: Scenario Preparation Program Database Development Guide, Jet Propulsion Laboratory, Pasadena, California, 1993.

Corps Battle Simulation: Scenario Preparation Program Language Manual, Jet Propulsion Laboratory, Pasadena, California, 1993.

Corps Battle Simulation: User's Handbook, Jet Propulsion Laboratory, Pasadena, California, 1993.

Dahmann, Judith S., "Command Forces Architecture", Presentation of the Advanced Research Projects Agency, 1994.

Defense Modeling and Simulation Office, "Panel Review of Aggregate Level Linkage Technologies", Aerospace Report Number ATR-92(2796)-1, The Aerospace Corporation, El Segundo, CA, February, 1993.

Dickson, Paul, *Think Tanks*, Ballentine Books, New York, NY, 1971.

"Distributed Interactive Simulation: Summary Report of the 10th Workshop for Interoperability of Defense Simulations", Orlando, Florida, March 1994.

Distributed Interactive Simulation: Operational Concept 2.3, UCF Institute for Simulation and Training, Orlando, Florida, September, 1993.

Distributed Interactive Simulation: Operational Concept, UCF Institute for Simulation and Training, Orlando, Florida, January, 1992.

Distributed Interactive Simulation: Standards Development Guidance Document, UCF Institute for Simulation and Training, Orlando, Florida, February, 1992.

DMSO Survey of Semi-Automated Forces, Defense Modeling and Simulation Office, March 15, 1993.

Economy, Richard, "Visualization Techniques—Putting the Reality in Virtual Reality", *Proceedings of the Virtual Reality Conference*, Data Processing Management Association, June, 1993.

Egdorf, H.W. and Painter, Steven W., "An Object-Oriented Methodology for Discrete-Event Simulation Tasks", Technical Paper, Los Alamos National Laboratory, Albuquerque, New Mexico, 1993.

Fishwick, Paul A., "A Simulation Environment for Multimodeling", Computer Science Technical Paper, University of Florida, 1994.

Fishwick, Paul A., "Computer Simulation: Growth Through Extension", Computer Science Technical Paper, University of Florida, 1994.

Fishwick, Paul A., "Utilizing abstraction and perspective in battle simulation", *Proceedings of the 1988 Winter Simulation Conference*, 1988.

Foster, Timothy A., "C2 Information Management: Data Fusion and Track ID's in a Multiple Sensor Environment", Masters Thesis, Naval Postgraduate School, March, 1992.

Franceschini, Robert W., "Intelligent Placement of Disaggregated Entities", Unpublished Paper, Institute for Simulation and Training, Orlando, Florida, 1994.

Fujimoto, Richard M., "Optimistic Approaches to Parallel Discrete Event Simulation", *Transactions of the Society for Computer Simulation*, Volume 7, Number 2, San Diego, CA, June, 1990.

Gehl, Thomas L., "Multicast Group Address Filtering for DIS", Position Paper, 9th DIS Workshop, September, 1993.

Geist, G.A., Beguelin, A., Dongarra, J.J., Jiang, W., Manchek, R., Moore, K., and Sunderam, V.S., *PVM 3 User's Guide and Reference Manual*, Oak Ridge National Laboratory, Oak Ridge, Tennessee, 1993.

General Dynamics, "Systems Analysis Models: JASMEAN Version 3.1 User's Manual", Fort Worth, Texas, 1989.

Genesereth, Michael R. and Ketchpel, Steven P., "Software Agents", *Communications of the ACM*, Vol. 37, No. 7, July, 1994.

Gibson, William, *Neuromancer*, Berkley Publishing Group, New York, NY, 1984.

Gilmer, J. B., "Parallel Simulation Techniques for Military Problems", Unpublished Paper, BDM Corporation, October 14, 1986.

Goodman, J.P., "A General Theory for the Fusion of Data", *1987 Tri-Service Data Fusion Symposium Technical Proceedings*, Johns Hopkins University, Laurel, Maryland, June, 1987.

Halsall, Fred, *Data Communications, Computer Networks, and Open Systems*, Addison-Wesley, Reading, Massachusetts, 1992.

Hardy, Doug, "BBS/SIMNET Functional Validation Test Report", Technical Report, Naval Command, Control and Ocean Surveillance Center, RDT&E Division, No Date, Approx. 1992.

Hardy, D.R., Healy, M., Owen, W., Jacobs, R., and Crooks, H., "Intelligent Gateway/Smart Switch (IGSS): BBS/SIMNET Integration Approach", Technical Document 2273, Naval Command, Control and Ocean Surveillance Center, RDT&E Division, January, 1992.

Haut, David G. and McCurdy, Michael L., "Modeling and Simulation in the New Pacific Community: A USPACOM Perspective", *Military Operations Research*, Summer, 1994.

Hughes Wayne P., "Uncertainty in Combat", *Military Operations Research*, Summer, 1994.

"Impact of Advanced Distributed Simulation on Readiness, Training, and Prototyping", Office of the Under Secretary of Defense for Acquisition, January, 1993.

Jacobson, Karie, editor, *Simulations: 15 Tales of Virtual Reality*, Citadel Press, New York, NY, 1993.

Jefferson, David, "Virtual Time", *ACM Transcations on Programming Languages and Systems*, Vol. 7, No. 3, July, 1985.

Jefferson, David and Sowizral, Henry, "Fast concurrent simulation using the time warp mechanism", *Distributed Simulation*, Vol. 15, No. 2, Society for Computer Simulation, 1985.

Johnson, George, *Machinery of the Mind*, Times Books, New York, NY, 1986.

Johnson, David B.,"Efficient Transparent Optimistic Rollback Recovery for Distributed Application Programs", Technical Report CMU-CS-93-127, Carnegie Mellon University, 1993.

Jones, Harold L. and Kronenfeld, Jerrold E., editors, "State-of-the-Art for WARSIM 2000", TASC Technical Report, TR-6947-3, Reading, MA, 1993.

Karr, Clark, "Integrated Eagle/BDS-D Interface Report", Institute for Simulation and Training, Orlando, Florida, January, 1994.

Karr, Clark, "Integrated Eagle/BDS-D Interim Report 3", Institute for Simulation and Training, Orlando, Florida, January, 1994.

Karr, Clark R. and Root, Eric, "Eagle/DIS Interface", *Proceedings of the 1994 Electronic Conference on Constructive Training Simulation*, listserv@mystech.com, 1994.

Kaudel, Fred J., "A Literature Survey on Distributed Discrete Event Simulation", ACM Simuletter, Vol. 18, No. 2, June, 1987.

Kochan, Stephen G., Wood Patrick H., Editors, *UNIX Networking*, Hayden Books, Carmel, IN, 1989.

Landry, L.M., "Charter for the Modeling and Simulation Industry Steering Group", Lockheed, Fort Worth, Texas, October 22, 1993.

Law, Averill M. and Kelton, W. David, *Simulation Modeling and Analysis*, McGraw-Hill Book Company, New York, NY, 1991.

Lee, Jin Joo, Norris, William Dean, and Fishwick, Paul A., "An Object-Oriented Multimodel Approach to Integrated Planning, Intelligent Control, and Simulation", Computer Science Technical Paper, University of Florida, 1994.

Levine, R.Y., "Neural Net Sensor Fusion", Technical Report 926, MIT Lincoln Laboratory, September, 1991.

Lin, Kuo-Chi and Ng, Huat, "Coordinate Transformations in Distributed Interactive Simulation (DIS)", *Simulation*, The Society for Computer Simulation, San Diego, California, November, 1993.

Mastaglio, Thomas W. and Rozman, Thomas R., "Expanding Training Horizons", *Army*, Association of the United States Army, February, 1994.

McLuhan, Marshall, *Understanding Media: The Extensions of Man*, Penguin Books, New York, NY, 1964.

Modular Semi-Automated Forces: Developer's Kit, Developed by Loral Advanced Distributed Simulation, Distributed by Tactical Warfare Simulation & Technology Information Analysis Center, Orlando, Florida, 1994.

Moshell, J. Michael, Blau, Brian, Li, Xin, and Lisle, Curtis, "Dynamic Terrain", *Simulation*, The Society for Computer Simulation, San Diego, California, January, 1994.

Othling, William and Speir, Les, *Analysis of Digital Topographic Data Requirements for Selected Army Models/Simulations. Volume I: Model Evaluations*, U.S. Army Topographic Engineering Center, Fort Belvoir, VA, 30 November 1993.

Payne, James A., *Introduction to Simulation: Programming Techniques and Methods of Analysis*, McGraw-Hill Book Company, New York, NY, 1982.

Pickett, Kent H., "DIS Architectural Elements for Tactical Representation of Higher Echelon Computer Generated Forces", Position Paper, 9th DIS Workshop, September, 1993.

Pimentel, Ken and Teixeira, Kevin, *Virtual Reality: through the new looking glass*, Windcrest Books, New York, NY, 1993.

Pooch, Udo and Wall, James, *Discrete Event Simulation*, CRC Press, Boca Raton, FL, 1992.

Popken, Douglas A., "Hierarchical Modeling and Process Aggregation in Object-Oriented Simulation", *International Journal in Computer Simulation*, Volume 4 Number 1, 1994.

Powell, Dave, "Distributed Interactive Simulation Architecture Components for the Battlefield Distributed Simulation—Developmental System", Position Paper, 9th DIS Workshop, September, 1993.

Proceedings of the Fourth Conference on Computer Generated Forces and Behavioral Representation, US Army Simulation, Training, and Instrumentation Command, Orlando, Florida, May 1994.

Protocol Data Units for Entity Information and Entity Interaction in a Distributed Interactive Simulation, Institute for Simulation and Training, Orlando, Florida, 1991

Rationale Document: Entity Information and Entity Interaction in a Distributed Interactive Simulation, Institute for Simulation and Training, Orlando, Florida, January, 1992.

"Real-Time Simulation Networking", Information Publication, Concurrent Computer Corporation, 1993.

Reilly, W. Scott and Bates, Joseph, "Building Emotional Agents", Technical Report CMU-CS-92-143, Carnegie-Mellon University, Pittsburgh, PA, 1992.

Reynolds, Paul F., "Early Performance Analysis and Recommendations: Reforger92", Presentation to TACSIM Project Office, October, 1992.

Reynolds, Paul F., "Parallel Simulation", Mystech Parallel Computing Workshop, August, 1993.

Reynolds, Paul F., "Reforger92: Performance Evaluation and Recommendations", Report to the U.S. Army TACSIM Project Office, October, 1992.

Rheingold, Howard, *The Virtual Community: Homesteading on the Electronic Frontier*, Addisson-Wesley, Reading, Massachusetts, 1993.

Ruck, Dennis William, "Characterization of Multilayer Perceptrons and their Application to Multisensor Automatic Target Detection", Unpublished Dissertation, Air Force Institute of Technology, Wright Patterson Air Force Base, Ohio, December 1990.

Salisbury, Marnie R., "Command and Control Simulation Interface Language", Presentation of the Advanced Research Projects Agency, 1994.

Schriber, Thomas J., *An Introduction to Simulation Using GPSS/H*, John Wiley & Sons, New York, NY 1991.

Seidensticker, Steve, "Distributed Simulation: A View from the Future", Unpublished paper distributed to DIS electronic mailing list, March, 1994.

Seidensticker, Steve and Lawler, Norm, "Balancing Act: A New PDU to Reduce DIS Network Traffic", 9th DIS Workshop, Orlando, FL, September, 1993.

Shirley, John, "Writing DCE Programs", *Unix Review*, January 1994.
Singley, George T., "Distributed Interactive Simulation—A Preview", *Army Research, Development, and Acquisition Bulletin*, March-April, 1993.

Siskind, Ron, "Concepts for Hex-free Play in CBS", Unpublished Paper, Jet Propulsion Laboratory, February 20, 1991.

"Software Design Document for the Advanced Interface Unit for the BBS-DIS Interconnect Computer Software Configuration Item", ETA Technologies Corp. for Naval Research and Development, August, 1992.

"Software Requirements Specification for the DOD Interlynx Network Interface (Strawman)", DIS Interface Subgroup, Orlando, FL, January, 1994.

Smith, Roger D., "Analytical Computer Simulation of a Complete Battlefield Environment", *Simulation*, 1992.

Smith, Roger D., "Applications of Virtual Reality Technologies to Existing Battlefield Simulations", *Signal*, 1993.

Smith, Roger D., "Applications of a Universal Data Analysis and Repository System in Military Simulations", *SCS Simulation MultiConference Proceedings*, 1993.

Smith, Roger D., *Battlefield Simulation: Analytical Techniques and Solutions*, ISA Press, Washington D.C., 1991.

Smith, Roger D., "JMEM Methodology in the ALBAM Simulation", Unpublished Technical Paper, General Dynamics, Fort Worth, Texas, 1990.

Smith, Roger D., "Multi-Simulation Software Interfaces", *Software Technology Conference Proceedings*, 1991.

Smith, Roger D., Notes from the 9th DIS Workshop, Orlando, Florida, September, 1993.

Smith, Roger D., editor, *Proceeding of the 1994 Electronic Conference on Constructive Training Simulation*, Falls Church, Virginia, May, 1994.

BIBLIOGRAPHY 163

Smith, Roger D., "TACSIM: Automated Intelligence Analysis and Enhanced Sensor Models", Symposium on Intelligence Applications of Modeling and Simulation, Fort Meade, Maryland, 1993.

Smith, Roger D., "Virtual Reality Technologies Integrated with Military Combat Simulations", *Software Review*, 1993.

Smith, Roger D. and Reynolds, Paul F., "Simulation Interfacing Techniques Proposal", Submitted to U.S. Army Materiel Command SBIR Office, Mystech Associates, August 1993.

Smith, Roger D., "Virtual Reality and Current Military Simulations", *Virtual Reality World*, 1994.

Software Developer's Document for the TACSIM After Action Review Users System, U.S. Army STRICOM, Orlando, Florida, 1993.

Software User/Operator's Manual for the TACSIM Relay for Input/Output Product, U.S. Army STRICOM, Orlando, Florida, 1993.

Somerville, Robert M. S., *The Military Frontier*, Time-Life Books, Alexandria, Virginia, 1988.

Standard for Distributed Interactive Simulation - Application Protocols, Version 2.0, Fourth Draft, UCF Institute for Simulation and Training, Orlando, Florida, February 14, 1994.

Steinman, Jeff S., "Breathing Time Warp", Unpublished Paper, Jet Propulsion Laboratory, Pasadena, California, 1993.

Steinman, Jeff S., "Incremental State Saving in SPEEDES Using C++", Unpublished Paper, Jet Propulsion Laboratory, Pasadena, California, 1993.

Steinman, Jeff S., "Parallel Proximity Detection and the Distribution List Algorithm", Unpublished Paper, Jet Propulsion Laboratory, Pasadena, California, 1993.

Steinman, Jeff S., "SPEEDES: A Multiple-Synchronization Environment for Parallel Discrete-Event Simulation", *International Journal in Computer Simulation*, Volume 2, 1992.

Strawman Distributed Interactive Simulation Architecture Description Document, Volume I: Summary Description, Loral Systems Company, Orlando, Florida, 31 March 1992.

Strawman Distributed Interactive Simulation Architecture Description Document, Volume II: Supporting Rationale, Book II: DIS Architecture Issues, Loral Systems Company, Orlando, Florida, 31 March 1992.

Strawman Distributed Interactive Simulation Architecture Description Document, Volume II: Supporting Rationale, Book I: Time/Space Coherence and Interoperability, Loral Systems Company, Orlando, Florida, 31 March 1992.

Summary Report: The Eighth Workshop on Standards for Interoperability of Defense Simulations, Volume I, II, and III, Institute for Simulation and Training, Orlando, Florida, 1993.

Summary Report: The Ninth Workshop on Standards for Interoperability of Defense Simulations, Volume I, II, and III, Institute for Simulation and Training, Orlando, Florida, 1993.

Summary Report: The Seventh Workshop on Standards for the Interoperability of Defense Simulations, Institute for Simulation and Training, Volume I and II, Institute for Simulation and Training, Orlando, Florida, 1992.

Summary Report: The Sixth Workshop on Standards for the Interoperability of Defense Simulations, Institute for Simulation and Training, Volume I and II, Institute for Simulation and Training, Orlando, Florida, 1992.

TALON User's Manual, U.S. Army STRICOM, Orlando, Florida, 1994.

TAARUS User's Manual, U.S. Army STRICOM, Orlando, Florida, 1994.

TACSIM-TALON/TAARUS Interface Controllers Document, U.S. Army STRICOM, Orlando, Florida, 1994.

TACSIM Liaison Officer's Manual, U.S. Army STRICOM, Orlando, Florida, 1994.

TACSIM Software Description Volume I: Functional Overview and Software Design, U.S. Army STRICOM, Orlando, Florida, 1991.

TALON-TAARUS System Administrator's Manual, U.S. Army STRICOM, Orlando, Florida, 1994.

Tanenbaum, Andrew S., *Modern Operating Systems*, Prentice Hall, Englewood Cliffs, NJ, 1992.

The DIS Vision: A Map to the Future of Distributed Simulation, Institute for Simulation and Training, Orlando, Florida, January, 1993.

U.S. Army Simulation, Training, and Instrumentation Command, "A Conceptual Architecture for Constructive Tactical Engagement Simulation: The WARSIM 2000 System", WARSIM 2000 White Paper, Paideia, September, 1992.

U.S. Army Simulation, Training, and Instrumentation Command, "Acoustic Environment Simulation for DIS Applications", WARSIM 2000 White Paper, BBN Systems and Technology, September, 1992.

U.S. Army Simulation, Training, and Instrumentation Command, "After Action Reviews: Keystone to Training", WARSIM 2000 White Paper, Lockheed Missiles & Space Company—Austin, September, 1992.

U.S. Army Simulation, Training, and Instrumentation Command, "Self-Describing Entities in a DIS Environment", WARSIM 2000 White Paper, BBN Systems and Technology, September, 1992.

U.S. Army Simulation, Training, and Instrumentation Command, "Simulation of Battlefield Radio Communications", WARSIM 2000 White Paper, BBN Systems and Technology, September, 1992.

U.S. Army Simulation, Training, and Instrumentation Command, "Tactical Communications for WARSIM 2000", WARSIM 2000 White Paper, GTE Government Systems, September, 1992.

U.S. Army Simulation, Training, and Instrumentation Command, "The Seamless Integration of Military Equipment into a Virtual Environment Simulator", WARSIM 2000 White Paper, BBN Systems and Technology, September, 1992.

U.S. Army Simulation, Training, and Instrumentation Command, "Use of Protocol Converters in WARSIM 2000", WARSIM 2000 White Paper, Lockheed Sanders, September, 1992.

U.S. Army Simulation, Training, and Instrumentation Command, "Variable Resolution Semi Automated Forces", WARSIM 2000 White Paper, BBN Systems and Technology, September, 1992.

U.S. Army Simulation, Training, and Instrumentation Command, "Voice Technology in WARSIM 2000", WARSIM 2000 White Paper, Lockheed Sanders, September, 1992.

U.S. Army Simulation, Training, and Instrumentation Command, "WARSIM 2000 and Semi-Automated Forces (SAFOR), A Rapid Prototype Approach", WARSIM 2000 White Paper, Martin Systems, September, 1992.

U.S. Army Simulation, Training, and Instrumentation Command, "WARSIM 2000 Architecture Discussion", WARSIM 2000 White Paper, Lockheed Sanders, September, 1992.

U.S. Army Simulation, Training, and Instrumentation Command, "WARSIM 2000 Concept Document", WARSIM 2000 White Paper, SAIC—Orlando, September, 1992.

U.S. Army Simulation, Training, and Instrumentation Command, "WARSIM 2000 Concept White Paper", WARSIM 2000 White Paper, IBM Federal Systems Corporation,September, 1992.

U.S. Army Simulation, Training, and Instrumentation Command, "WARSIM 2000: Flexible and Predictable Performance in Distributed Simulation", WARSIM 2000 White Paper, Lockheed Missiles & Space Company—Austin, September, 1992.

U.S. Army Simulation, Training, and Instrumentation Command, "WARSIM 2000 White Paper", WARSIM 2000 White Paper, Digital Equipment Corporation, September, 1992.

U.S. Army Simulation, Training, and Instrumentation Command, "WARSIM 2000 White Paper", WARSIM 2000 White Paper, GE Advanced Technology Laboratories, September, 1992.

Vakili, Pirooz, "Massively Parallel and Distributed Simulation of a Class of Discrete Event Systems: A different Perspective", *ACM Transactions on Modeling and Computer Simulation*, Volume 2 Number 3, New York, NY, July, 1992.

Vasend, Gerald, "Intelligent ALSP/DIS Gateway", Unpublished Paper, Logicon/RDA, 1994.

Weatherly, Richard; Seidel, David; and Weissman, Jon, "Aggregate Level Simulation Protocol", MITRE Informal Report, McLean, Virginia, July 1991.

West, Phillip D., "NPSNET: Object Animation Script Interpretation System", Unpublished Thesis, Naval Postgraduate School, Monterey, California, September, 1991.

Woolley, Benjamin, *Virtual Worlds*, Blackwell Publishers, Cambridge, Massachusetts, 1992

Yourdon, Edward, *Decline & Fall of the American Programmer*, Yourdon Press, Englewood Cliffs, New Jersey, 1992.

Zeigler, Bernard P., *Theory of Modelling and Simulation*, Robert E. Kreiger Publishing Company, Malabar, Florida, 1984.

Appendix A—Acronyms

1 1 0 0 1 0 0 0 1 1 0 0 1 0 1 0 0 0 1 1 0 0 1 0 1 0 1 0 1 0 0 0 1 1 0 0 1 0 1 1 1 0 0 1 1 1 0 0 1 0 1 0

AAA	After Action Analysis (1)
AAA	Anti-Aircraft Artillery (2)
ABE	ALSP Broadcast Emulator
ACM	ALSP Communication Module
ADCATT	Air Defense Combined Arms Tactical Trainer
AI	Artificial Intelligence
AIUBBS	Adapter Interface Unit—Brigade/Battalion Battle Simulation
AIUSAF	Adapter Interface Unit—Semi Automated Forces
ALBAM	Air Land Battle Assessment Model
ALLRAD	All Range Air Defense
ALSP	Aggregate Level Simulation Protocol
ARPA	Advanced Research Projects Agency
ASF	Advanced Simulation Framework
ATD	Advanced Technology Demonstrator
ATM	Asynchronous Transfer Mode
AVCATT	Aviation Combined Arms Tactical Trainer
AWSIM	Air Warfare Simulation
BBN	Bolt, Bereneck, & Neuman Corporation
BBS	Brigade/Battalion Battle Simulation
BDS-D	Battlefield Distributed Simulation—Developmental
BICM	Battlefield Intelligence Collection Model
BN	Battalion
CAC2	Combined Arms Command and Control
CATT	Combined Arms Tactical Trainer
CBS	Corps Battle Simulation
CCTT	Close Combat Tactical Trainer

CDR	Commander
CFOR	Command Forces
CGF	Computer Generated Forces
CGS	Command Ground Station
CID	Combat Identification
CIS	Combat Instruction Sets
CPU	Central Processing Unit
CSSTSS	Combat Service Support Training Simulation System
DARPA	Defense Advanced Research Projects Agency
DEC	Digital Equipment Corporation
DIS	Distributed Interactive Simulation
DMSO	Defense Modeling and Simulation Office
DOD	Department of Defense
DR	Dead Reckoning
ENCATT	Engineering Combined Arms Tactical Trainer
FACATT	Field Artillery Combined Arms Tactical Trainer
FDDI	Full Duplex Data Interchange
GOTS	Government Off The Shelf
GPS	Global Positioning System
GRWSIM	Ground Warfare Simulation
GVT	Global Virtual Time
HIMAD	High-Medium Range Air Defense
HY-DY	High Dynamics
IEEE	Institute for Electrical and Electronic Engineers
IFOR	Intelligent Forces
IST	Institute for Simulation and Training
JECEWSI	Joint Electronic Combat / Electronic Warfare Simulation
JMEM	Joint Munitions Effectiveness Models
JPL	Jet Propulsion Laboratory
LAN	Local Area Network
LOSAT	Line-Of-Sight Anti-Tank
LP	Logical Processor
MAIS	Mobile Automated Instrumentation Suite
MAN	Metropolitan Area Network

MI	Master Interface
MIL	Man-In-The-Loop
MIPS	Millions of Instructions per Second
ModSAF	Modular Semi-Automated Forces
MS-DOS	Microsoft Digital Operating System
NASA	National Air and Space Administration
NPS	Naval Postgraduate School
NTC	National Training Center
OI	Operator Interface
OSI	Open Systems Internet
PDU	Protocol Data Unit
PK	Probability of Kill
PO	Persistent Object
PO-DB	Persistent Object—Database
RASPUTIN	Rapid Scenario Preparation Unit for Intelligence
RESA	Research and Analysis Simulation
RF	Radio Frequency
RPA	Rotorcraft Pilot's Associate
RPC	Remote Procedure Call
SAF	Semi-Automated Forces
SAFOR	Semi-Automated Forces
SAM	Surface-to-Air Missile
SHORAD	Short Range Air Defense
SIMCON	Simulation Controller
SIMD	Single Instruction Multiple Device
SIMNET	Simulator Networking
STOW	Synthetic Theater of War
STRICOM	Simulation, Training, and Instrumentation Command (U.S. Army)
TAARUS	TACSIM After Action Review Users System
TACSIM	Tactical Simulation System
TALON	TACSIM Analysis Operations Node
TCTS	Tactical Combat Training System
TOC	Tactical Operations Center

TRAC	TRADOC Analysis Center (U.S. Army)
TRADOC	Training and Doctrine Command (U.S. Army)
TRIOP	TACSIM Resource for Input/Output Processing
TSAM	Tactical Surface-to-Air Missile Model
UTM	Universal Transverse Mercater
VAX	Virtual Access Extensions
VLIV	TACSIM Simulation Engine
WAN	Wide Area Network

Appendix B—Glossary

1 1 0 0 1 0 0 0 1 1 0 0 1 0 1 0 0 0 1 1 0 0 1 0 1 0 1 0 1 0 0 0 1 1 0 0 1 0 1 1 1 0 0 1 1 1 0 0 1 0 1 0

Anti-Aircraft Artillery: Rapid-firing machine guns, usually equipped with radar tracking systems, capable of shooting down aircraft.

Aggregate Level Simulation Protocol: A family of software modules and communication protocols designed to join multiple constructive simulations in a single shared exercise.

ALSP Broadcast Emulator: ALSP Component used to replicate the operations performed by the TCP network broadcast.

ALSP Communication Module: Component used to connect a constructive simulation to the ALSP network.

Air Defense Combined Arms Tactical Trainer: Planned system for training the crews of air defense batteries.

All Range Air Defense: Air defense missiles used at for long, medium, and short range targets.

Aviation Combined Arms Tactical Trainer: Planned system for training the members of an aviation maintenance crew.

Combined Arms Command and Control: Planned system for training members of the command staff in virtual level simulations.

Combined Arms Tactical Trainer: Armor vehicle crew training simulation system under development by Loral Federal System Company and team mates.

Computer Generated Forces: Virtual simulation system which allows a single operator to control multiple vehicles in a realistic manner. The operator is assisted by knowledge-bases and artificial intelligence algorithms embedded in the system.

Combat Instruction Sets: An effort designed to capture the essential elements of command decisions and relationships in a structured manner. This information will be used to extend the power of Computer Generated Forces systems.

Distributed Interactive Simulation: A concept and set of protocols intended to join multiple virtual simulations in a single shared environment.

Engineering Combined Arms Tactical Trainer: Planned system to train the members of an engineering platoon vehicle operators. This includes those that create and remove minefields, build and destroy bridges, and create and overcome barriers.

Field Artillery Combined Arms Tactical Trainer: Planned system to train the members of an artillery battery.

Global Positioning System: Network of satellites which broadcast location signals. These signals are collected and triangulated by receivers which then provide an accurate location of the vehicle using the receiver.

Global Virtual Time: Concept which allows parallel simulation processes to operate on different local simulation clocks. GVT provides methods for synchronizing events that occur between parallel processes.

High-Medium Range Air Defense: Air defense systems which use missiles to engage targets at both high and medium ranges.

Intelligent Forces: Conceptual simulation models which can replicate the operations of humans, learn from combat experiences, and modify their behavior appropriately.

Institute for Simulation and Training: Research branch of the University of Central Florida. Its primary mission is to support simulation concepts originating from the U.S. Army Simulation, Instrumentation, and Training Command in Orlando, Florida.

Man-In-The-Loop: A class of simulation in which human operators plan an interactive part.

Modular Semi-Automated Forces: Computer generated forces system being developed to support multiple simulation projects needing such a device. Software and documentation are distributed by the Institute for Simulation and Training.

Protocol Data Unit: The information packets used by the Distributed Interactive Simulation system to create an integrated synthetic combat environment.

Reforger: The largest military exercise in Europe prior to its demise following the end of the cold war. Designed to prepare NATO forces for a Soviet invasion of western Europe, which would require a Return of Forces to Germany (Reforger).

Semi-Automated Forces: Simulations systems and algorithms designed to allow a single human operator to control multiple vehicles at the virtual level.

Short Range Air Defense: Air defense systems used to engage targets at short ranges.

Simulator Networking: Prototype system developed by DARPA to explore the feasibility of creating large networks of simulators which are able to participate in a single shared synthetic environment.

Synthetic Theater of War: The army's concept of a battlefield which completely exists in computer memory.

Universal Transverse Mercator: A method for measuring location on the earth using Cartesian coordinates. Though not globally continuous like Latitude/Longitude, it provides a simple systems for determining accurate locations. Primarily used by the Army.

www.ingramcontent.com/pod-product-compliance
Lightning Source LLC
Chambersburg PA
CBHW021052210326
41598CB00016B/1186